Robotic Process Automation in Use

Mario Richard Smeets · Andreas Freßmann ·
Ralf Jürgen Ostendorf

Robotic Process Automation in Use

Strategic Orientation—Practical Implementation—Audit-Proof Implementation

Springer

Mario Richard Smeets
Viersen, Germany

Andreas Freßmann
Harsewinkel-Greffen, Germany

Ralf Jürgen Ostendorf
Krefeld, Germany

ISBN 978-3-658-45904-8 ISBN 978-3-658-45902-4 (eBook)
https://doi.org/10.1007/978-3-658-45902-4

This book is a translation of the original German edition “Robotic Process Automation im Einsatz” by Mario Richard Smeets et al., published by Springer Fachmedien Wiesbaden GmbH in 2023. The translation was done with the help of an artificial intelligence machine translation tool. A subsequent human revision was done primarily in terms of content, so that the book will read stylistically differently from a conventional translation. Springer Nature works continuously to further the development of tools for the production of books and on the related technologies to support the authors.

This Springer imprint is published by the registered company Springer Fachmedien Wiesbaden GmbH, part of Springer Nature.
The registered company address is: Abraham-Lincoln-Str. 46, 65189 Wiesbaden, Germany

Contents

About the Authors

Mario Richard Smeets helps his clients as a management consultant to streamline processes and automate them using the latest technologies such as RPA and Artificial Intelligence. As the managing partner of a consulting firm specializing in process automation, he supports corporate executives in their strategic challenges in dealing with new technologies, but also in their implementation. Mario Smeets is a Master of Business Administration with a focus on Management of Financial Institutions and a Master of Science in Economics.

Dr. Ralf Jürgen Ostendorf is a Professor of Finance and Business Management at the Niederrhein University of Applied Sciences and a lecturer at the Osnabrück University of Applied Sciences—Department for Dual Studies. His main areas of research, teaching, and practice are: financing, investment, organization, controlling, as well as the strategic orientation of institutions and individuals. He gained practical experience, among other things, as a Prokurist (authorized officer) and head of controlling at BAG Bankaktiengesellschaft in Hamm, as well as a manager in responsible positions at the Sparkasse in Sprockhövel and the MOHAG mbH Recklinghausen. The author holds the following academic degrees: Dipl.-Ök., Dipl. Bankbetriebswirt ADG, Dipl.-Hdl. and Dipl.-Soz.-Wiss.

(LichtBlick—Photography by Paul Wiesmann, Recklinghausen)

Andreas Freßmann is the Head of Audit and Authorized Officer of Volksbank Beckum-Lippstadt eG. He has more than twenty years of audit and leadership experience in the savings bank and cooperative sector in all areas of banking audit work (sales, operations, credit, and IT audit). The author is a graduate in business administration.

Introduction 1

In recent years, the use of software-based tools for process automation has become established, particularly in medium to large companies. There, the support is primarily in administrative areas, predominantly in the back offices and related business areas. These tools, often technically very extensive, have also increasingly found their way into banks. Here, they are also predominantly used in the back offices (active and passive), but also in the IT or controlling area and in accounting. There, tables are filled, data is transferred from one application to another, or digital documents are checked. These processes regularly take place completely rule-based and following rigid patterns.

The automation of such tasks creates space for employees to perform other, more complex tasks. Until a few years ago, the almost exclusive solution for automating such processes was the creation of interfaces or the (re-) programming of the affected applications—all usually technically only feasible with high effort or not at all, as well as cost- and time-intensive. The tools predominantly used today can be summarized under the term "Robotic Process Automation—RPA"—they enable significantly more cost-effective automation than the conventional methods used until a few years ago (van der Aalst, 2022).

With RPA, applications can be automated without having to intervene in program codes or create interfaces. The McKinsey Global Institute attributed a (aggregated) technical automation potential of 43% to the global financial industry as early as 2017, which is to be tapped piece by piece in the coming years and decades (McKinsey Global Institute, 2017). Even if verifiable figures for a status comparison as of today are missing at this level of survey, the core statement is clear: The financial industry and especially banks, which traditionally still use many non-automated processes, have enormous potential for efficiency increases and cost savings through automation, most of which can be tapped with RPA. Even though many banks have already recognized this potential and are using RPA in initial projects or for smaller process areas, the number of companies with

M. R. Smeets et al., *Robotic Process Automation in Use*,
https://doi.org/10.1007/978-3-658-45902-4_1

a comprehensive use of the technology is still low, albeit increasing (Smeets et al., 2021).

The trend towards process automation in banks is supported by an increasing change in the relevant competitive environment. For example, a—at least currently still ongoing—low interest rate situation leads to potentially sustainably lower interest income, (process) costs are rising due to regulatory requirements or innovations in consumer protection. These external effects sometimes require strategic reorientations of banks in their competitive environment. RPA as a tool for process automation can support this (re-) positioning.

Against the background of an already increasing intensity of use of RPA technology in banks and the possibly even further increasing degree of increase, it can be assumed that RPA will leave or has already left the stage of "technology testing" in many places. At the latest from this point, the technology also becomes relevant for comprehensive auditing and supervision by the internal (or even external) audit. Due to the still low maturity level of the technology (or its use in banks), there is a lack of comprehensive information, conditions and concepts that would allow an audit to enable a targeted and in every respect "safe" use of RPA within their own bank.

After an introduction to RPA technology and a demarcation from other similar technologies, the potential of RPA in the context of the mentioned strategic (re-) positioning is first explained. This is followed by an intensive examination of possible requirements for an audit-side accompaniment of the use of technology. This leads to conceptual considerations that enable readers to use RPA "audit-proof" and with low risk in their own institution. Here, audit staff, other audit bodies, but also project managers or other users of RPA (and possibly other automation technologies) will find many valuable information and support. Unlike comparable basic works on RPA in the financial sector (e.g. Smeets et al., 2021), the focus of the book is less on an intensive examination of basic thoughts on RPA technology. Rather, these are briefly discussed before illuminating other exciting and relevant aspects in the RPA or automation context with strategic considerations and audit-side aspects.

All chapters—(2) RPA—Background and Introduction, (3) RPA as a tool for strategic positioning in competition and (4) RPA from the perspective of the audit—can be read and used as independent chapters. Chronological reading is therefore not necessary, the reader's focus can immediately be placed on the individually preferred content.

References

McKinsey Global Institute. (2017). *A future that works: Automation, employment, and productivity*. McKinsey Global Institute.

Smeets, M., Erhard, R. U., Kaußler, T. (2021). Robotic Process Automation (RPA) in the financial sector. Technology—Implementation—Success for decision makers and users. Springer Gabler. http://www.springer.com/.

van der Aalst, W. M. P. (2022). Process mining and RPA: How to pick your automation battles?

2 Robotic Process Automation—Background and Introduction

2.1 Definition and Delimitation of RPA

In the literature, there are various definitions of RPA. The most striking differences in the definitions lie in the delimitation to similar, self-learning technologies that use artificial intelligence. This is a source of frequent misunderstandings, which are also repeatedly found in practical discussions. First, a clear definition of RPA will be given, based on Smeets et al. 2021a. This is followed by a further delimitation of RPA from other, similar technologies.

Definition RPA or the term "robot" does not refer to a physical machine, but to installable software. It supports humans in carrying out their (digital) activities and processes, acting as a kind of virtual assistant. In most cases, processes can be automated with RPA when they are rule-based and structured. RPA communicates with other applications and systems, controls them, and transfers or manipulates data. This requires no or only minor interventions in existing IT architectures, as RPA usually uses the user interface to operate the automated applications. Interventions in the program code or the use of interfaces to the automated applications are not necessary. The software robots can work both on the desktops of employees and in a kind of dark processing, e.g., on servers.

In contrast to other works (e.g., Ranerup & Henriksen, 2019), a delimitation of RPA in its pure form and Artificial Intelligence (AI) takes place here, as RPA does not use independent learning or similar, which in turn is an essential part of the definition of AI (see e.g., Kolbjørnsrud et al., 2017; Mahroof, 2019; Jarrahi, 2018).

Figure 2.1 shows this delimitation from other technologies. On the vertical scale, the degree of artificial intelligence that a tool possesses is plotted; on the horizontal scale, the achievable degree of automation. This relationship is easy to explain. In its pure form, RPA is initially only suitable for the automation of

M. R. Smeets et al., *Robotic Process Automation in Use*,
https://doi.org/10.1007/978-3-658-45902-4_2

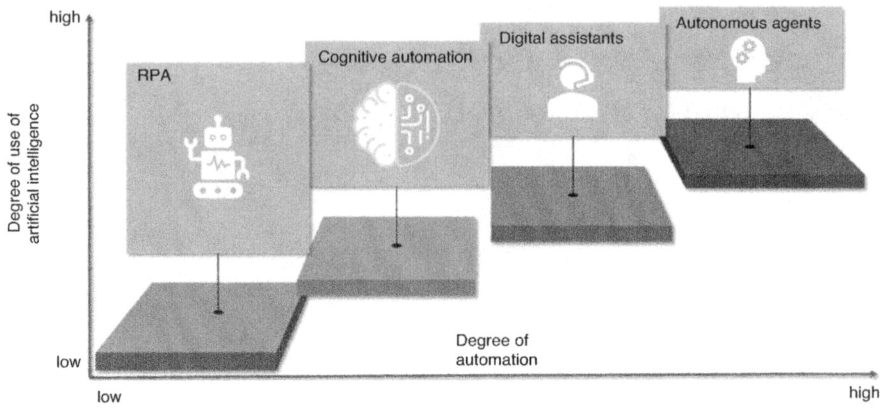

Fig. 2.1 Delimitation of RPA and artificial intelligence. (Based on Ostrowicz, 2018)

completely rule-based processes. No decisions can be made that are not (manually) predefined. The further to the right in the graphic, the more and more complex decisions can be made by the tool. Accordingly, more processes are potentially suitable for automation than with a purely rule-based solution (or larger parts of a process can be automated, especially in complex, long processes). The figure is to be understood schematically, but it helps with a quick classification of RPA in the context of automation technologies.

The RPA tools initially considered here have no machine learning capabilities or properties of artificial intelligence. The latter can be understood as further developments of existing and tested RPA solutions. The "Automation Journey" of the RPA-using institute usually begins with RPA (Ostrowicz, 2018). The degree of artificial intelligence is low (by definition even zero); accordingly, the spectrum of possible applications is limited—the number of processes that can be automated is relatively small.

> **Important** A study conducted by the authors in the savings bank sector revealed a proportion of approx. 20–25% of processes that can be automated within a "typical" savings bank; the basis is the standard process map. Although the quantity is relatively small compared to the quantities achievable with other technologies, it is still considerable when viewed absolutely.

"Cognitive Automation"—as a further technology—has machine learning capabilities in contrast to RPA and recognizes patterns within small amounts of unstructured data. So here we can already speak of "artificial intelligence". Cognitive methods are used in particular for structuring previously unstructured content. "Digital Assistants" or "Social Robots" (Allweyer, 2016) even interact directly with employees and/or customers. They also have the ability to recognize patterns.

However, the data input can be even more unstructured, for example as text or speech. Both technologies use technologies such as Natural Language Processing (NLP) or Optical Character Recognition (OCR) in addition to machine learning. "Autonomous Agents" represent the (currently) highest level of development. They can automate highly complex tasks by simulating human judgment using mathematical models. Comprehensive market maturity was—at least until the end of the last decade—not yet given (Ostrowicz, 2018).

All four outlined solutions have different areas of application. Autonomous agents aim at the analysis of large, unstructured data sets, so work is done in the background (dark processing). The focus of digital assistants lies in communication with internal or external customers, so not in the background, but in direct interaction. RPA, on the other hand, automates repetitive, standardized processes with high volumes (Smeets et al., 2021a). The solutions thus complement each other and together form a kind of "toolbox" that can be used for so-called hyperautomation (Jalali-Sohi, 2021; Gartner, 2022).

Definition Hyperautomation refers to a rapid and comprehensive automation of many processes in a company through the combined use of several technologies. These include, for example, RPA, machine learning (i.e., artificial intelligence), business process management (BPM), or process mining (see also Gartner, 2022). Simplified and exemplified (without being able to go into details here), process mining can help identify relevant processes that are then automated with RPA and artificial intelligence. Subsequently, BPM is used as a tool for controlling the entire workflow (workflow management).

A possible process flow that can be automated through hyperautomation is shown in Fig. 2.2.

In the outlined process, a message from the outside comes in, for example, a customer message. The machine-unreadable and unstructured text is first made machine-readable using optical character recognition (OCR), then structured using artificial intelligence, and can then be further processed with RPA.

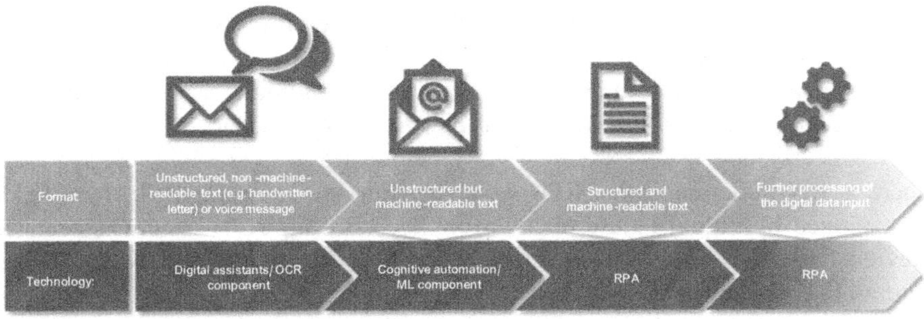

Fig. 2.2 Fully automated process using combined technologies—Hyperautomation. (Smeets et al., 2021a)

> **Important** The RPA tools used in practice today (as of 2022) increasingly contain pre-installed cognitive components. For example, an OCR component is standard, as are initial possibilities for recognizing structures in unstructured data. It can therefore be assumed that today we are no longer talking about RPA in its pure form, but rather RPA with cognitive components ("Cognitive RPA"). This will also be the case in the following—not only because of its practical relevance, but also because many important questions arise from this with regard to accompanying audits.

2.2 RPA—An Introductory Technical View

The following chapter outlines some relevant technical peculiarities of RPA. The aim is to provide an overview of the structure of the technology, its possible integration into the existing IT architecture of a bank, and its communication with other applications. Detailed insights into individual software or programming instructions are not relevant here.

2.2.1 Applications Automatable with RPA

In principle, all applications—i.e., all software—that have a user interface can be automated with the common RPA tools. Exceptions cannot be found in the literature, and the authors are not aware of any cases from practice (unless a software provider deliberately blocks the application against the use of technical tools on their surface; as seen in a single case so far in an application of a provider of market data for the trading departments of banks).

Even applications without a direct user interface are usually automatable. Here, RPA can access via (usually standardized) interfaces. This can then be used to work directly in databases, etc.

> **Important** Even with applications that are typically operated via the user interface, the alternative use of interfaces may be advisable. Accesses via these are usually (still) more stable than via the surface, as they are protected against unexpected changes. Updates/changes usually only occur here after announcement and are rarer with standard systems. In addition, access is faster, as latency and loading times of the user interfaces do not occur.

There is still the opinion that RPA would operate "screen scraping". Here, fields are operated by virtual mouse click/virtual keyboard input. These clicks occur at previously defined places on the screen. So the bot is given the instruction to execute a click at coordinate xy. The problem: If the element to be operated moves, the bot can no longer select or edit it, or in the worst case makes incorrect entries.

Important Recommendation: In any case, a connectivity check between RPA software and the applications to be automated should be carried out before setting up an RPA project or licensing specific software. The software providers or IT consulting firms are usually available for uncomplicated trials.

2.2.2 Comparison of RPA and BPM

The previous section explained that RPA primarily accesses applications via the user interface; direct, interface-based communication often takes place with databases. When accessing via the user interface, the systems to be automated do not recognize that they are being operated by another software. They react as if a human were operating them and making entries (Smeets et al., 2021a). This is different with conventional automation solutions, which can be found in Business Process Management. These communicate with the database or the application via the database access layer. Fig. 2.3 illustrates the comparison.

While access to business logic and database access layers often need to be laboriously programmed, access via the user interface represents a simple and fast access path. RPA thus provides a flexible automation solution independent of the specific requirements of the target application.

2.2.3 Structure of RPA Solutions

RPA tools consist of different components for design, control, and execution. The common providers use at least three:

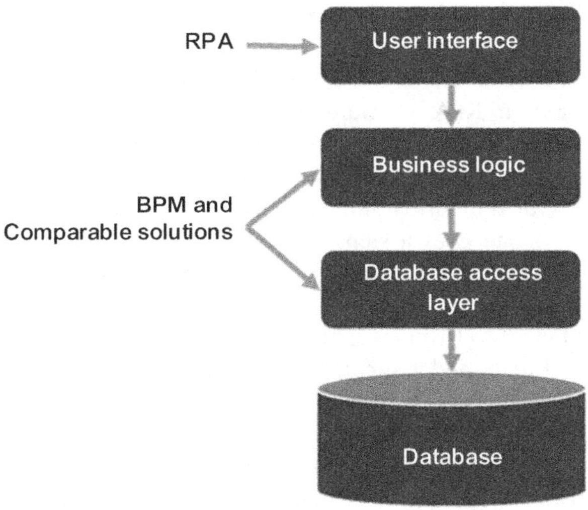

Fig. 2.3 Database access by RPA and BPM and comparable solutions. (Based on Lacity & Willcocks, 2016)

One component to implement process automation, i.e., to "design" the process flow for the RPA process. Here, we deliberately do not speak of "programming", as RPA process design is more akin to assembling in a modular system, at least in its simple variant. This is done with a high degree of graphical support. Flowcharts describing process flows in a form known from process management are created using "drag-and-drop". Selection lists provide predefined commands, so no or only minimal programming effort is necessary. Many tools have a recording function that records the processes in their basic structure and creates the basic framework for further processing by RPA developers.

The second relevant component is the actual "bot", i.e., the executing instance. This executes the created RPA processes (also "RPA artifacts"). It is thus the component that—figuratively speaking—replaces the process-executing human (Smeets et al., 2021a). Here it applies: A bot can execute any number of different processes. However, this is sequential, not parallel. For example, the bot executes several hundred transfers at the start of the day (Process 1), then makes changes to customer master data for one to two hours (Process 2), before processing the next batch of transfers (Process 1). As soon as the bot reaches its capacity, another instance, i.e., another bot, needs to be licensed.

Important Although bots can technically be used 24/7, the runtimes, especially in banks, are often limited by end-of-day processing.

The more bots are operated, the more important the third component becomes, the central control unit, often also referred to as the "Orchestrator". With it, the individual bots can be controlled and capacities optimally distributed. This is where utilization and performance are monitored, but also updates and patches are centrally installed and distributed to all executing instances.

2.2.4 RPA Architecture

The installation and use of RPA software is seemingly simple at first glance. Modern RPA software is considered infrastructure-independent (Smeets et al., 2021a). It can be operated on the desktops of employees (also "attended" RPA), or server-based as "real" dark processing (also "unattended" RPA). RPA can also be operated in virtual infrastructures—in private or public cloud environments.

If one goes a step further, a (technical) evaluation of the integration of RPA with regard to the bank's own IT architecture is necessary. It is advisable to immediately consider all guidelines—be they internal or regulatory, for example those of the BAIT (this will be discussed in more detail later).

Banks and their applications and systems usually have three infrastructural environments: A development, a test, and a production environment. The latter is where live operation takes place. When implementing RPA, these environments can be used: The development of RPA processes takes place in the development environment, the subsequent testing in the (often closer to production) test environment. Only after a final acceptance test does the "go-live" in the production environment follow. Especially with RPA, a tool that uses the user interfaces of the applications, a high identity of all environments is necessary. Otherwise, errors or different behavior in production operation may occur.

2.3 Goals and Potentials of RPA Use

In order to understand and recognize possible RPA use cases, it is necessary to discuss the different goals or potentials that a bank can pursue and leverage with the use of RPA.

> **Important** At least four categories can be distinguished: cost savings, quality improvements, time savings, and other potentials, which often can be derived from the first three (here and in the following Smeets et al., 2021a, b).

2.3.1 Cost Savings

One of the most frequently and often first mentioned arguments for process automation are the possible cost savings that can be achieved. These are possible primarily because employees can use their capacities for other tasks while the RPA bots completely take over repetitive processes or handle parts of them. This can result in a reduction of the bound capacities, also FTE (Full Time Equivalents), for the respective process. However, it should be emphasized that automation by RPA does not necessarily mean that the personnel previously commissioned for the execution of the process are released and no longer incur costs. Rather, the aim is to free employees from repetitive, time-consuming processes through RPA, so that more capacities are available for value-adding activities, such as creative work. The extent of potential cost savings through RPA varies and depends on the process (its complexity and many other factors), the application area, and the industry. The cost savings range from 25% (see, for example, Watson & Wright, 2017) to 90% (Smeets et al., 2021a). The focus—in the literature—is typically in the range of about 30% to a maximum of 70%. The magnitudes also differ depending on the cost factors included. License costs, maintenance costs, and more should be considered, which unfortunately is not always the case. Based on the authors' many years of practical experience with RPA, a magnitude of about 25% cost savings compared to the current process with RPA technology seems to be a realistic value for the financial industry, especially when all cost items are taken into account.

2.3.2 Quality

In general, the principle probably applies everywhere that the quality of a work result should be as high as possible. However, this assumption, i.e., the quality goal, is not always explicitly formulated as a strategic goal and is often rather taken for granted. However, the quality improvement of a process or a process result can indeed be defined as an explicit strategic goal for an RPA deployment. From the authors' point of view, it is even the greatest benefit gained from the use of RPA. Because: When humans carry out processes, this is always prone to errors, especially with frequent, repetitive tasks. The process execution by RPA bots, however, is not—provided the processes (more precisely: the technical artifacts) are correctly developed and sufficiently tested. Unsystematic errors—i.e., errors that occur randomly, e.g., caused by humans—are excluded by RPA. The only risk that remains is that of systematic errors, which should not be neglected. A faulty programming that is only noticed at the introduction of RPA can quickly lead to large amounts of incorrect process results. This can be counteracted with appropriate tests, which are also demanded by regulatory authorities in the financial industry for software development (in Germany, for example, the already mentioned and later detailed BAIT).

2.3.3 Time Savings

Time savings or reduction of processing time is the possible third, overarching goal of automation with RPA. Time reduction usually goes hand in hand with cost savings, as it is often the case for process costs that they are reduced by shorter processing times. The possible time savings are nevertheless worth being considered as a separate strategic goal. Especially in processes where the (external) customer is directly or indirectly involved, an increase in speed can generate competitive advantages and improve customer satisfaction. This applies not only to external customers, i.e., customers of the institute. Also internal areas that are involved in processes or trigger them can be considered as customers and as recipients of the services of another area also have the expectation of a high process speed. Here, RPA contributes significantly to an increase in process speed.

If the goal is "time savings", the following things are important and should be considered: The speed of the robot depends largely on the underlying process. If this has been appropriately prepared and optimized ("process-streamlining") before automation, the bot can usually work faster. The quality of process development also influences its speed. The response times of automated systems cannot be influenced. Since the robot works on the surface (the graphical user interface—GUI), it is always dependent on the speed of these applications.

2.3.4 Further Potentials

The three main objectives or most important potential areas of RPA mentioned above cover a large part of the benefits argument for RPA. From this, further potentials can be derived. For example, the technology can be used to reduce compliance and similar risks. Here, two possibilities are conceivable: Once an RPA process has been developed according to the specifications and protected against change, it can no longer be unintentionally and unnoticedly changed. This significantly limits possible misuse. In addition, every step and every system input of the bot is documented, both within the target applications and in the RPA's own documentation. This means that all actions are fully verifiable by third parties, such as auditors or financial auditors. The same applies to operational risks. These include human errors, system errors, or errors within internal processes. Operational risks are a significant risk category, especially in the financial sector (Kaiser & Koehne, 2007). The discussed increase in process quality when using RPA reduces these risks. Secondly, it is possible to carry out significantly more checks, tests, etc. with RPA than with humans. Standardized test routines or preparatory activities such as information gathering, which is particularly necessary in the audit areas, can be quickly and easily automated with RPA. The latter is also an exciting use case for the audit departments themselves, which are explicitly part of the target audience of this book.

In practice, a high workload for the IT department is regularly observed. This is particularly the case when the bank has a heterogeneous system landscape with many applications and legacy systems. Here too, RPA can be used supportively by offering an alternative to building interfaces between these applications and systems.

RPA offers further potentials for the financial industry by shortening the time-to-market for the placement of new products or the establishment of new processes, by the standardized recording and evaluation of information, and much more.

All these potentials make the "in-house processing" of processes, or even the insourcing of already outsourced processes, attractive again, especially from a cost-benefit perspective. The use of RPA also raises questions about outsourcing and outsourcing management. In the further course, questions related to this will also be discussed.

2.4 Areas of Application for RPA

2.4.1 Main Areas of Application

As has been mentioned several times, almost any process can be automated with RPA. This is at least technically true—the situation is different when additional business economic criteria are taken into account. Here, it is critical and process-specific to question whether automation has a positive cost-benefit ratio, i.e., whether the savings exceed the costs of automation and ongoing operation. When such criteria are considered, the process clusters with the best cost-benefit ratio can be found particularly in the following areas (Ostrowicz, 2018; Smeets et al., 2021a).

Primary Areas:

- Back-Office
- Finance (for example, accounting)
- IT

In addition, the following areas also have a basic technical and business economic potential for RPA (secondary areas):

- Compliance
- Legal
- Risk or overall bank management
- Human Resources

The mentioned areas can be derived from the authors' years of experience in automating a three-digit number of processes. In comparison with further assessments of the relevant literature, the estimates vary.

For example, a study by the Information Services Group (ISG) assumes that IT will be most affected by automation in the long term (Otto & Longo, 2017). Other studies deviate from this, for example, a survey conducted by PWC in 2017 among several dozen professionals in the USA. According to this study, the back-office areas offer the greatest potential with well over 80% approval. Finance and IT follow. In the areas of compliance, legal, risk controlling, and human resources, only a small potential for RPA is seen, the same applies to sales. Ostrowicz (2017) confirms the applicability of RPA in back-office, risk management, accounting, and HR (Smeets et al., 2021b).

> **Important** Wherever data is available in a digital format and is processed digitally, the use of RPA is recommended.

If one now adds the degree of standardization of a process area as a criterion, it becomes clear why the process areas of the back office and, for example, accounting are suitable for RPA. Here, most processes are standardized, rule-based, and use digitized data. The typical sales areas, on the other hand, are less suitable for automation (at least with little effort) according to the understanding of RPA used here. In the sales area, the use of desktop RPA is also quite conceivable, as the authors have already been able to accompany the successful use of RPA as a support tool for sales staff in (customer) data entry on several occasions. Here, existing processes often have to be broken down into individual sub-processes, which can then be partially automated.

2.4.2 Roles of RPA

In order to be able to present possible use cases in a structured manner, the three different roles that RPA can take on must first be defined (summarized by Smeets et al., 2021b):

1. RPA as support for humans in the execution of existing processes
2. RPA as a replacement for humans in the execution of existing processes
3. RPA as a possibility to establish completely new business models or processes

The classic role of RPA is that of a digital assistant. The bots support humans in completing their tasks or provide relevant recommendations for decisions. These types of bots often work on the desktops of employees, are started and stopped by them, and interact with them. It should be noted that the bots operate in the applications of the employees and often use their access data. This means that employees often cannot make entries themselves while the bot is working. A relevant distinguishing feature to "unattended" RPA, i.e., the previously explained dark processing of data.

The second role that RPA can take on is a complete replacement of humans or employees. Here, RPA carries out certain activities or individual tasks independently and without human intervention. While in the first role a human sets the trigger for the start (or end) of the bot's work, for example by giving it a start and end signal, the bot in the second case is started by technical triggers. Alternatively, flow charts or similar can be stored for controlling the robot. These are then executed and controlled by the central control unit (the previously mentioned orchestrator).

In both of the above cases, it is assumed that the bot is used to automate existing processes. This is usually the case, especially at the beginning of RPA use in a company. Here, simple and rule-based processes are used and automated in a first step. But once RPA is established, it offers opportunities to set up new processes at high speed and in partly new ways. If new market requirements are placed on products or if the legal framework conditions change, new processes can often be necessary or make sense from a sales and business management perspective. So far, these could often only be composed of many manual activities, as technical adjustments were too expensive or took too long. RPA now offers the possibility to set up new processes quickly and flexibly—even if technical interfaces and other tools are not available.

For the following use cases see Smeets et al. 2021a, b as well as practical experiences of the authors.

2.4.3 Use Cases of RPA as a Support Tool

Trade

In the trading areas of a bank, a multitude of cross-application automation possibilities can be listed. Traders continuously monitor course information or news. If these come from a single source, often trading terminals, further automation with RPA is usually not necessary. However, as soon as information from other systems or, for example, from third-party websites is to be observed and evaluated, flexible automation with RPA is recommended. The bots monitor market prices and information on websites and send information to the traders according to predefined patterns or when certain events occur. The advantage of RPA lies in the integration of different data sources and the linking of different systems that do not originally have interfaces with each other.

Provision of Regular Reports

The controlling department of a bank, often organizationally assigned to overall bank management, regularly creates a multitude of reports and evaluations. Various information sources from usually different systems and applications are used. In addition to internal bank databases, content from various websites is also accessed, to which no interfaces can be created. These can be, for example, company data from competitors that are used for benchmarking. As part of automation, the bot can condense such content and data and merge it in a structured form. This content is then made available to the bank's management in a consistently identical layout. The employees then take over the quality assurance, carry out the final evaluation steps, and finalize the reporting or the report. Depending on the quality of the RPA results and depending on the respective process, it is even conceivable that humans are completely replaced by bots. However, this is initially a support service provided by RPA.

2.4.4 Use Cases for RPA as a Substitute

Invoice Input and Output Processing

Incoming invoices are usually not booked immediately. Before the invoice amount is transferred, a check is made for correctness and legality of the invoice. If the necessary data for this is available in digital form, immediate automation is possible. For this, RPA can be used in its pure form. However, if the data is still in an unstructured form, an OCR component must be added to make the data processable by RPA. This is the case when invoices are received by letter—a clear decreasing trend and a shift towards the medium of email can now be observed.

IT—Focus on Authorization Management and Passwords

When hiring employees, with every department change, every competence change and every employee departure, various permissions must be granted, changed or

deleted. In many cases, these are similar "permission packages", especially when hiring new employees or when employees leave the company. An ideal use case for RPA, with the help of which employee permissions can be managed in all necessary applications, so that human intervention is only necessary for basic specifications such as master data or the areas of responsibility to be assigned.
Another example is the resetting of passwords; also daily practice. Here too, resetting by bots is easily possible—provided digital (and standardized) input channels are chosen. Forms or ticket systems are suitable here—emails or even phone calls from employees should be avoided in favor of data structure.

RPA for One-Time Use Cases
In most cases, the use of RPAs aims at the permanent automation of processes, i.e. the use of technology in day-to-day business. In addition, RPA is also suitable for one-time use cases, such as data migrations. If a technical transfer of the data is not possible, often the only option is to use employees: time-consuming and—due to the manual work shares—prone to errors. After mapping old and target data, RPA can often offer a cheaper, faster, but above all error-free alternative. The use of RPA has recently also proven itself in the implementation of system updates or in the processing of large amounts of data. Here too, RPA is the tool of choice when alternative technical solutions cannot be used.

Checking of Securities Settlements and Other Bookings
The reconciliation of trading transactions is still often done manually in many places. Often several employees are tied up in this every day. Individual transactions are recorded in internal applications and transferred to spreadsheet programs for reconciliation. At the same time, visual checks and comparisons are carried out. All routines that can be almost completely automated. The advantage of using RPA here lies in the usually complete digitality of the data.

2.4.5 Use Cases for RPA in Establishing New Processes

Digital Product Conclusions
Product conclusions are now increasingly made online, including in banks. While the entry of customer data is regularly already digital, the further processing is often still done manually. Interfaces between the customer frontend and the core banking system are not always available. RPA can form or replace this interface. In the target image, bots transfer the data recorded by the customer into the legally leading system—after checking for completeness and plausibility. This achieves a true end-to-end digitization of the process, and the processing time and thus the costs of the individual processes remain low.

Compliance
The compliance department, like other control bodies of the institutions, monitors and checks various processes and other activities. For example, suspicious

transactions on accounts are checked. Usually, only random checks can be carried out, as complete monitoring is not possible in terms of resources. With RPA, much more extensive test routines can be set up, even complete monitoring is possible on a case-by-case basis, if desired.
The use cases listed up to this point are often examples from the authors' practice. In addition, many other examples could be listed.

2.5 Selection of RPA-Suitable Processes

2.5.1 Selection Criteria

The selection of RPA-suitable processes can be based on partially standardized, but also customizable criteria. Basically, a distinction is made between technical and business criteria (Table 2.1).

> **Warning** Only a few processes are completely automatable. According to a relevant study, less than 5% of all processes are completely automatable across all sectors, 95% accordingly not or only in parts. But even in these 95%, there is enormous potential, because about 60% of the activities considered here have a share of automatable process components of about 30% (McKinsey Global Institute, 2017).

Therefore, when selecting a process, it is important not to interpret the technical criteria too strictly, especially at the beginning of the selection process. A process that contains non-automatable steps can still be a sensibly (partially) automatable process. An appropriate division of the process is necessary here. Examples are shown in Fig. 2.4.

In the first variant, a migration process is outlined. Data is extracted from the old application, processed, and then inserted into the target application. All steps are digital and standardized, based on fixed rules. Here, full automation is possible. The situation is different in the second process example, a sales process. Here, the RPA bot assists in preparing the customer conversation, but the actual conversation is conducted by a human. Afterwards, the bot takes over again and processes all data captured (and thus digitized) by the human.

2.5.2 Procedure for Process Selection

When a company is dealing with the introduction of RPA for the first time, a potential target process for piloting is usually already known. Often these are processes that bother many employees due to their low complexity (often a "blunt processing" is possible) and high case numbers, thus tying up a lot of capacity. Although already known, the process selection should be questioned at least once

Table 2.1 Selection criteria RPA processes. (Smeets et al., 2021a)

Criterion	Technical/Business	Remark
Degree of standardization	Technical criterion	The more standardized, the better automatable.
Rule-based	Technical criterion	Complete rule-based = complete automation possible, otherwise check partial automation.
Process stability/maturity	Technical criterion	The more stable the process (the fewer expected changes), the more suitable it is.
Complexity	Technical criterion	The lower the complexity, the more suitable it is.
Data type	Technical criterion	Text and numbers are suitable, but images or handwritten data are less so.
Involved applications	Technical criterion	The more applications the process goes through and the higher the number of system breaks, the more sensible automation with RPA, but the complexity also increases (counteracting effect)
(Digitality of the data)	Technical criterion	Only digital data can be processed by the bot. Considering Cognitive RPA, this criterion may be omitted (e.g., with OCR component)
(Structure of the data)	Technical criterion	RPA in its pure form can only process structured data, i.e., those that the bot receives in the previously expected form. This criterion can also be omitted with Cognitive RPA.
Case frequency	Business criterion	The greater the case frequency, the greater the potential for cost savings through automation.
Process costs	Business criterion	The higher the actual process costs, the more worthwhile automation is.
Error susceptibility	Business criterion	The greater the error susceptibility of a manual process, the more likely RPA can lead to quality improvements.
Process processing time	Business criterion	The longer the actual process processing time, the more worthwhile automation is, as the cost saving is potentially higher.

more within the framework of a professional RPA pilot—based on the criteria explained above and the procedure for process selection outlined below. As soon as further processes are to be automated with RPA after the automation of the first process, a structured approach to the selection of the processes in question is necessary, as has been proven in numerous implementation projects in practice.

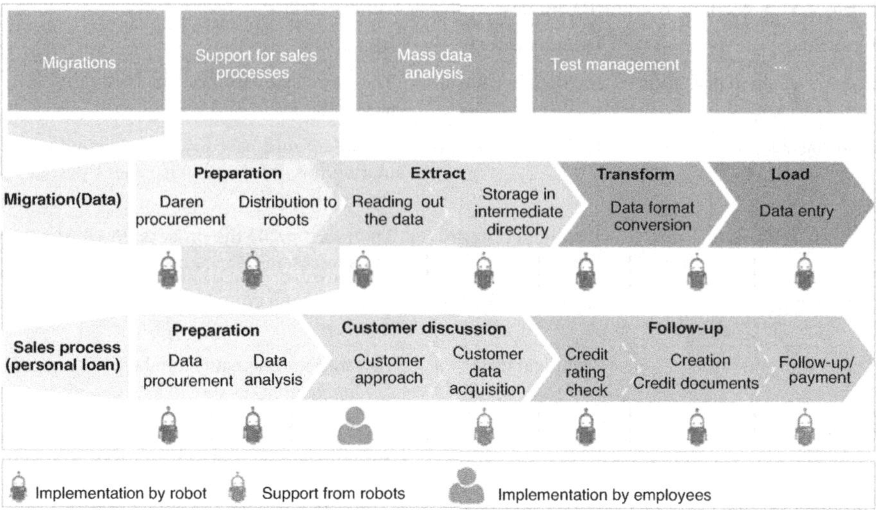

Fig. 2.4 Partially and fully automatable processes (own illustration)

Firstly, three basic approaches to process selection can be distinguished:

1. Current process
2. Expert workshops
3. Process map

In the first case, the process, as initially described, is already known. In addition to the "pilot processes" often in the focus of employees, certain processes may also be explicitly intended for automation in the further course of RPA use in the company. These can also be processes that do not meet the explained criteria (especially not the business ones), but for which there are other, individual reasons for automation—risk aspects, quality improvement efforts, etc.

Example

An RPA deployment accompanied by the authors involved data migration between two core banking systems. The establishment of a technical interface and thus a machine migration was not possible in terms of time. The initially only alternative: manual data entry by a large number of employees. This would have entailed major quality risks due to erroneous entries. The solution was RPA: quickly set up and thanks to correct programming (which had been extensively and structured tested during the project, like a typical IT development) with error-free migration performance. In the end, not only was the quality goal achieved, but a cost saving of (in relation to other RPA automations "only") almost 30% was realized, as the use of additional temporary workers had become obsolete. ◄

The second possibility of a structured process selection is the conduct of expert workshops. Here, for example, workshops are held with individual departments or departments that show capacity problems or want to save costs, in which the relevant selection criteria are first explained and possibly supplemented by individual ones, and then processes of the employees are jointly examined and prioritized based on the criteria. A possible workshop document for process selection is shown in Fig. 2.5 and comes from the authors' consulting practice.

In the rows, individual processes can be collected and evaluated using the criteria in the columns. Especially when conducting for the first time or at the beginning of an RPA deployment, it is advisable to have such workshops accompanied by (external) RPA specialists. The existing know-how can then be internalized piece by piece and over time.

The third possibility involves analyzing the complete process inventory of an institution or a department. It is the most complex and time-consuming. However, it allows the potential for RPA in the individual institution to be determined accurately. The depth of detail of such an analysis can be decided individually, but the result should at least be a statement about each process, whether it is fully, partially, or not automatable. Such analyses are usually also carried out in practice with external support.

▶ **Important** Regardless of the choice of the basic approach, a three-step procedure for evaluating the individual processes has established itself in practice. In a first step, all potential processes are evaluated based on a few technical selection criteria. The aim is to determine whether a process is automatable or not. In the second step, the remaining processes are examined in more detail, still with a technical focus. Here, an assessment of complexity and possible technical challenges that make automation more difficult compared to others is made. In the third step, the remaining processes are evaluated from a business perspective. Here, benefits, usually monetary savings, and estimated automation costs are compared. In addition to this process-specific analysis, the individual processes can also be compared with each other in this step. It is advisable to use the net benefit of the processes as a comparison criterion. This corresponds to the benefit (see above, usually cost savings) minus the estimated automation costs.

2.6 Implementation of RPA

The following chapter describes the procedure for implementing RPA. It is assumed that this is done on a project basis. Often, initial uses of RPA in the company start with a project-based trial of the new technology. Subsequently, follow-up projects are regularly initiated until a transition to the line occurs (Smeets et al., 2021a). From this point on, project-independent activities can be observed in practice, in particular, the maintenance of automated processes is carried out just like the setting up of new processes in the "daily business".

	Units involved					Technical criteria			Business economist. Criteria											
process	Cust	Consultant	Assistance	Operations	25 limit?	Degree of digital data input	Degree Rule-based/structured	Degree of stability	ACTUAL proces duration (short 5/long 1)	Quantity Process runs (high 5/low 1)	Quality improvement potential	Meaning Process speed keti	Score	Rank according to criteria	From which process step digital?	Process executable as "batch"?	Attended or unattended	Process runs per day	Process runs p.a.	Average processing times
process 1	x	x	x	x	Yes	4	3	4	5	4	2	5	27	1	Customer data acquisition	no	at	5	1250	60
process 2		x	x	x	Yes	1	2	1	2	2	3	3	14	3	Start of trial	Yes	un	10	2500	5
process 3				x	no	5	5	2	4	4	3	3	26	2	Start of trial	Yes	un	10	2500	10
													0	4					0	
													0	4					0	
													0	4					0	
													0	4					0	
													0	4					0	
													0	4					0	

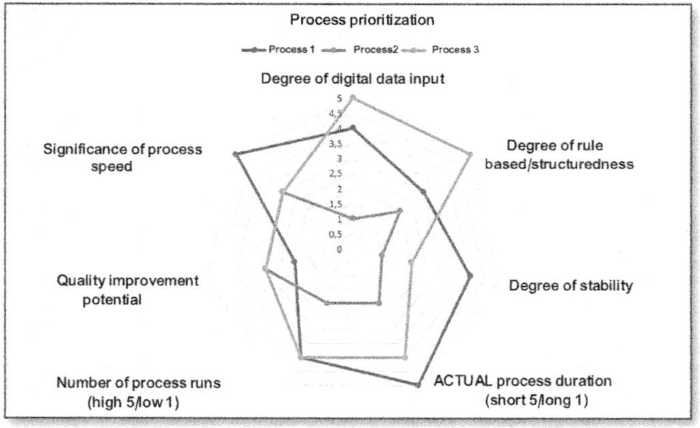

Fig. 2.5 Process selection matrix (own representation)

The advantages of a project-based implementation of RPA are in particular (see also Smeets et al., 2021a):

- Sufficient budget, time and resources
- High attention of potential stakeholders
- "Green field" to gain experience in dealing with RPA
- Risk minimizing, as initial piloting is possible
- "Careful" introduction of the technology to stakeholders and employees

Important An 8-step approach is recommended, which is shown in Fig. 2.6. The procedure outlined there has proven itself in practice. It ensures a structured project process and prevents forgetting individual, relevant steps.

Section 4.3.3.3.5 describes individual project steps from the perspective of audit. The focus there is on the requirements of the BAIT. The procedure described here takes these regulatory requirements into account within the individual steps.

2.6.1 Setting up project structure

As in any project, setting up a suitable project structure is the starting point. Orders/goals are to be clearly named, responsibilities defined and budgets distributed. In the case of an RPA project, it is advisable to involve both IT staff and the departments whose process/processes are initially the target of automation. Other relevant stakeholders for a project in which a new technology is being tested are regularly the audit and compliance area.

If not already done, it is advisable to involve the necessary RPA specialists—usually from outside—when setting up the project structure. At least one process and RPA/AI business analyst and one RPA/AI developer are needed. The team

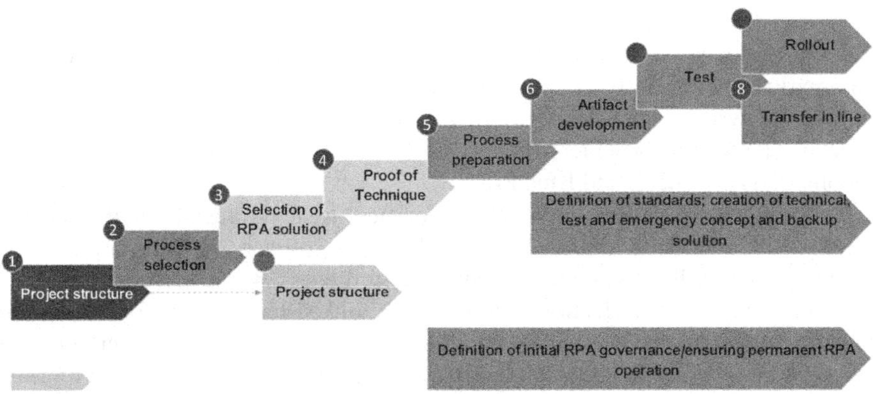

Fig. 2.6 Project-based implementation of RPA. (Smeets et al., 2021a)

size depends on the scope of the project. Practice shows that usually one employee capacity (MAK) per role is sufficient.

In practice, it has proven useful not to staff RPA projects exclusively externally, but to define internal project staff to at least the same extent (see also Smeets et al., 2021a). These should in particular be those employees who will later also be responsible for the technology or work with it in daily operation. The project to be set up can be divided into individual subprojects. Here too, the necessity depends on the overall size of the project. However, at least the topics mentioned must be taken into account. The individual subprojects or topic areas are:

- The organizational implementation of RPA, i.e., the "anchoring" within the later line operation
- The technical implementation of RPA, i.e., the installation of the software on servers and clients
- A process selection, analysis and possibly adjustment
- The actual development/process automation
- A testing
- The final rollout into productive operation

2.6.2 Process selection

The second step is the selection of the processes that are to be automated as part of the (pilot) project. This is always necessary when these are not already determined before the implementation of the automation. The basic procedure for process selection was outlined in the previous chapter. It is always advisable when processes are to be selected and prioritized in workshops, or when entire process maps are to be examined for automatable processes (Smeets et al., 2021a).

2.6.3 Selection of the RPA Solution

If not yet selected, a decision for a provider of RPA software needs to be made. The market is constantly changing, which is why a list of possible providers at the time of writing this chapter does not seem sensible. However, the currently three most relevant providers on the market can be named—these are UiPath, Automation Anywhere, and Blue Prism.

The software selection is of great relevance, as it is crucial for the automations of the following years. A switch between software—once started in the company—is hardly possible, each automated process would have to be set up/configured anew. The selection is also closely linked to the selection of the processes to be automated. Individual RPA solutions interact better with certain applications than others. Also, the simplicity of process development or the conformity with the existing IT architecture within the bank can be relevant distinguishing features of the individual solutions. The following list represents possible differentiation options (Smeets et al., 2021a).

- Introduction costs
- License costs
- Expected other costs
- Software components and usability
- Support/Training
- Audit, information, and compliance security
- System requirements and integration capability
- Required additional services
- Provider's experience in the industry, considered process area, etc.
- Location, legal form, etc.
- Provider's support
- Product roadmap

2.6.4 Proof of Technique

Following the software selection—or even before a decision—a "Proof of Technique" (PoT) is recommended. The PoT refers to the technical testing of the favored RPA software. Here, the interaction with the applications to be automated is practiced. Points to be tested are—in addition to the aforementioned connectivity—the possibility of installing the RPA software in the in-house systems and environments as well as the reading and writing ability within all relevant applications.

Sometimes, PoTs are also carried out as part of the selection of the RPA software to be used. No matter when, a comprehensive PoT is important. A later change of the software brings massive challenges and regularly requires a lot of time. Not only the automations aimed at in the direct (pilot) project should be considered, but also possible later automations. These can be completely different applications, which should accordingly also be checked for their connectivity at the beginning. Often, these applications can be identified by identifying possible future target processes or business areas at the beginning of the automation journey.

2.6.5 Process Preparation

The fifth step involves preparing the process (or processes) to be automated. Often neglected in practice, the relevance of this preparation cannot be emphasized enough. A poor process that is automated will simply become a poor automated process. An impending automation can be a good starting point for a comprehensive process analysis and improvement. The steps of process preparation include in particular:

- The recording of the current process
- A sketch of the desired state
- The analysis of necessary adjustments to move from the current to the desired state
- The development of a concrete approach to implement the adjustments

Such process preparations regularly take place in workshops. Different individuals or groups of people should be involved. In addition to the RPA business analyst, ideally the RPA developer should participate directly in order to understand the later target process in detail. It is advisable to involve participants from both the IT and specialist sides in these process workshops. The IT side is the contact for all technical/application-related questions that arise in the process analysis or design of the target process. The specialist side provides the technical input, so it can, for example, provide information on whether certain process steps are actually necessary or can be omitted. In addition, the "process ownership" and thus the best knowledge regarding the relevant process is regularly located here.

The results of the workshops are recorded in so-called Process Definition Documents (PDD). These contain all relevant information, in particular (Smeets et al., 2021a):

- Overview page with all relevant information such as
  a) Creation date
  b) Revisions/adjustments
  c) Creating unit/person
  d) Process responsible unit/person
  e) Bot developer
  f) Possibly further individual information
- Representation of the overall process with identification of the individual subprocesses
- Individual page per subprocess with
  a) Previously created screenshots at the lowest level of detail
  b) Illustration of the process in workflow diagram form
  c) Verbal description of the steps
  d) Hints on special features
  e) References to additional data to be used (for example, tables outside the application to be automated that the bot should use)

The PDD not only serves the developer in implementing the automation. The process test can also be carried out using the document. In addition, it provides a basis for the review of the RPA implementation by the audit. Everything that is in the PDD must also be technically implemented—conversely, nothing may be implemented that is not documented in the PDD.

2.6.6 Artifact Development

Based on the PDD, the actual RPA process is technically "developed" or "configured" in the sixth step. This can be done in a waterfall or agile manner—recently, an agile approach has proven successful in practice. First, a so-called "MVP—Minimal Viable Product" is created, a term frequently used in agile development. Initially, the standard case is automated, i.e., the main branch of the process tree is mapped. Special cases are initially disregarded. In the second step, exceptions and side branches follow. Only in the third step are special cases also integrated. The scope here is correspondingly large, a multitude of possible process variants can be automated regularly. However, complete automation is rarely achieved, i.e., an automation degree of 100% is achieved. The reason for this is exceptions and special cases that occur in practice and are so individual and rare that the benefits of automation would not outweigh the associated costs.

RPA development is not software development in the usual sense. It can rather be spoken of a configuration, as programming is rarely or not at all done (no lines of code or similar are written). Nevertheless, it is advisable to consider relevant principles of software development. This includes, for example, the choice of the correct application environment. Thus, process development should take place within a development environment of the application to be automated, the subsequent test in the test environment, and only the live operation in the production environment. If there are no development and test environments available or if these deviate too much from the production environment, development in production operation can exceptionally take place. In this case, however, further precautions must be taken to ensure security, such as restricting user rights and comprehensive documentation. Regular contacts for the organization-specific requirements in the context of development are revision and information security management (Smeets et al., 2021a).

2.6.7 Test

The testing phase takes place either after development, or already in parallel. The latter especially when an agile development approach has been chosen or when the development of individual, self-contained process parts takes place. A three-step approach is recommended (see also Smeets et al., 2021a).

1. Creation of a suitable test concept
2. Test execution
3. Acceptance and release

In the first step, it is defined exactly what and to what extent should be tested. It is important to fully consider all activities carried out by the RPA bot. The process is therefore to be tested "end-to-end". In addition to the test scope, it is also necessary to determine during the conception who will carry out the tests, when they will be carried out, and who will confirm the successful test execution at the end.

It is advisable to consider the institution-specific guidelines for software development—if available. As a rule, this also ensures compliance with regulatory requirements (e.g., BAIT).

The second step is the execution of the tests. This does not mean the (ongoing) developer tests that the RPA developers carry out as part of setting up the RPA processes. Rather, the subsequent technical tests are meant, which ideally should be carried out by technically knowledgeable people, e.g., the process owners. All test runs must be documented comprehensively. This includes in particular a test case description, used test data, and the result of the test. Ideally, in addition to positive tests, i.e., tests where a positive result is expected, negative tests are also carried out. This means tests where an error is expected (e.g., "The bot must not accidentally trigger event xy, but must stop beforehand with an error").

The final acceptance and release certifies a positive conclusion of the RPA development. It only takes place when the results defined in the test concept are consistently achieved. In the case of a pure (rule-based working) RPA bot, this is usually a 100% result, i.e., the complete and always correct processing of the relevant process. If a component is integrated that is based on artificial intelligence (i.e., cognitive RPA is used), this can be deviated from. It is common to define percentage thresholds at which a functioning process is assumed. The erroneous cases are then to be processed by employees. In addition, it can be assumed that the number of errors decreases over time.

2.6.8 Rollout and Handover into the Line

If the tests have been successful and the acceptance has been certified, the rollout (or go-live) of the RPA process can take place. Several dimensions need to be considered here:

First, a technical transfer of the developed RPA process into the production environment is necessary. This ensures that all security requirements are met and a start of the bot is possible. In addition to the purely technical dimension, it is also necessary to adequately prepare the employees interacting with the new RPA process. This involves concepts or instructions for process execution and possibly even training. Employees who do not directly interact with the RPA process should ideally also be informed and prepared for the changed process world.

If multiple processes are going live, an order must be defined. The same applies if a process is to be used in several areas. In both cases, it often makes sense to act with low risk, i.e., to opt for sequential go-lives instead of parallelization.

After go-live, a short intensive support phase is recommended. During this phase, it should be closely monitored whether the RPA bot behaves as expected and tested in production operation. If not, quick adjustments need to be made here. It should be noted that "quick" adjustments in the production environment are generally not possible. Instead, errors must be corrected again in the development environment, tested in the test environment, accepted, and then only taken over as an adjusted process into production operation.

Once the go-live and the intensive support phase are completed, the institute has to decide how to proceed with the RPA technology. Often the project form is maintained and further processes are selected and automated. If sufficient training of the employees has already taken place, a handover of the RPA responsibility into the line operation can be considered. In this case, the support/maintenance of the automated process and the automation of further processes are carried out from the line unit.

2.6.9 Definition of Operational Standards and Establishment of RPA Governance

Parallel to the steps described here, it is recommended to define basic conditions for daily RPA operation from the beginning and also to make arrangements for dealing with technical failures.

Like any other application, the RPA software itself needs to be maintained and its operation needs to be supported. It must be ensured that suddenly occurring errors can be analyzed and corrected as quickly as possible. With extensive use of RPA, production support can bring corresponding challenges and must be planned. Estimates regarding the number of employees required per bot range from about 0.1 to 0.3. This means that one person supports about three to ten bots. The ratios are fundamentally dependent on a variety of factors, such as the know-how of the supporters, the maturity of RPA in the company, or the complexity and stability of the automated processes. The responsible employees should be so trained and have so much RPA know-how that they can recognize, analyze, and solve most of the technical problems themselves. This applies especially to the technical part. The employees do not have to be process owners or have technical knowledge about the automated process. Rather, they support the technical managers in solving the problems.

There are different approaches to the organizational classification of production support. It is often seen in the IT area, sometimes in the organizational area, and—rarely—in other areas, such as the technical area (Smeets et al., 2021a).

Other relevant areas that should be defined by an RPA governance are the following (Smeets et al., 2021a):

- The strategic goals of RPA use should be defined. It is recommended to align or derive them from the existing strategic corporate goals.
- Existing processes should be checked once and continuously for their automatability. This continuously identifies potentials. In addition, a regular process for identifying such potentials in ongoing operation should be established.
- Responsibilities and guidelines for managing automated processes and bots in operation need to be implemented.

- It is advisable to build up RPA know-how and to operate corresponding knowledge management in the company.
- Incentive systems can be created for the identification of automation potentials and similar.
- An often already existing "Continuous Improvement Process" should also be used in relation to RPA.

The organizational integration of RPA is also defined within the framework of RPA governance. This will be discussed in more detail in Chap. 4 .

2.7 Summary

Chapter 2 provided an overview of RPA technology and possible areas of application (especially in banks), as well as suggesting a possible approach to implementing the technology.

- The trend towards automation of previously manual processes is stable and is now also noticeable in many areas of the financial services sector
- Robotic Process Automation is a tool in the process manager's toolbox that can be used to automate (digital) processes simply, quickly, and without necessary technical interfaces
- RPA can be supplemented with additional components, such as Artificial Intelligence (AI)—this is then referred to as Cognitive Automation or Hyperautomation
- The main objectives of using RPA are usually "cost savings", "quality improvement" and "time savings", but other things—such as a fast time-to-market for new processes—can also be main objectives
- The most important areas of application for RPA are the back office, finance, and IT
- Suitable processes for potential exploitation are those that are digital, rule-based, and structured (if AI is added, the criteria "structured" and "rule-based" can be somewhat relaxed)
- The implementation of RPA follows an 8-step model and is based on the approach used in traditional software development

References

Allweyer, T. (2016). *Robotic Process Automation—Neue Perspektiven für die Prozessautomatisierung*. Working Paper Fachbe-reich Informatik und Mikrosystemtechnik Hochschule Kaiserslautern. http://www.kurzeprozesse.de/blog/wp-content/up-loads/2016/11/Neue-Perspektiven-durch-Robotic-Process-Automation.pdf. Accessed: 5. Sept. 2022.

Gartner. (2022). *Hyperautomation*. https://www.gartner.com/en/information-technology/glossary/hyperautomation. Accessed: 9. Sept. 2022.

Jalali-Sohi, M. (2021). *Hyperautomation, strategischer Technologietrend für Unternehmen.* Bigdata Insider. https://www.bigdata-insider.de/hyperautomation-strategischer-technologie-trend-fuer-unternehmen-a-1078308/. Accessed: 9. Sept. 2022.

Jarrahi, M. H. (2018). Artificial intelligence and the future of work: Human-AI symbiosis in organizational decision making. *Business Horizons, 61*(4), 577–586. https://doi.org/10.1016/j.bushor.2018.03.007.

Kaiser, T., & Koehne, M. F. (2007). *Operationelle Risiken in Finanzinstituten. Eine praxisorientierte Einführung* (2nd edn.). Gabler (SpringerLink Bücher).

Kolbjørnsrud, V., Amico, R., & Thomas, R. J. (2017). Partnering with AI: How organizations can win over skeptical managers. *Strategy & Leadership, 45*(1), 37–43. https://doi.org/10.1108/SL-12-2016-0085.

Lacity, M., & Willcocks, L. (2016). Robotic process automation at Telefónica O2. *MIS Q Exec, 15*(1), 21–35.

Mahroof, K. (2019). A human-centric perspective exploring the readiness towards smart warehousing: The case of a large retail distribution warehouse. *International Journal of Information Management, 45,* 176–190. https://doi.org/10.1016/j.ijinfomgt.2018.11.008.

McKinsey Global Institute. (2017). *A future that works: Automation, employment, and productivity.* McKinsey Global Institute.

Ostrowicz, S. (2017). *Einsatz von Robotics in der Finanzindustrie.* https://www.horvath-partners.com/es/media-center/studien/detail/einsatz-von-robotics-in-der-finanzindustrie/. Accessed: 24. Jan. 2019.

Ostrowicz, S. (2018). *Next Generation Process Automation: Integrierte Prozessautomation im Zeitalter der Digitalisierung. Ergebnisbericht Studie 2018.* Horváth & Partners.

Otto, S., & Longo, M. (2017). *ISG-Studie: Robotic Process Automation (RPA) sorgt für mehr Produktivität und nicht für Jobverluste.* https://www.isg-one.com/docs/default-source/default-document-library/isg-automation-index-de_final_form.pdf?sfvrsn=15defe31_0. Accessed: 20. Jan. 2019.

Ranerup, A., & Henriksen, H. Z. (2019). Value positions viewed through the lens of automated decision-making: The case of social services. *Government Information Quarterly, 36*(4), 101377. https://doi.org/10.1016/j.giq.2019.05.004.

Smeets, M., Erhard, R. U., & Kaußler, T. (2021a). *Robotic Process Automation (RPA) in the financial sector. Technology—Implementation—Success for decision makers and users.* Springer Gabler. http://www.springer.com/.

Smeets, M. R., Ostendorf, R. J., & Rötzel, P. G. (2021b). RPA for the financial industry. In C. Czarnecki & P. Fettke (Hrsg.), *Robotic process automation. Management, technology, applications* (pp. 263–284). Walter de Gruyter GmbH (De Gruyter STEM).

Watson, J., & Wright, D. (2017). *The robots are ready. Are you?*https://www2.deloitte.com/content/dam/Deloitte/tr/Documents/technology/deloitte-robots-are-ready.pdf. *Accessed: 9. Sept. 2022.*

3 RPA as a Tool for Strategic Positioning in Competition

After the second chapter focused on the advantages and possibilities of RPA deployment, a step to the side now follows: the presentation of the current competitive situation in the banking sector. This is followed by the preparation of traditional competitive strategies, including their reflection for the banking sector. Based on these results, a further development of potentially promising orientations in competition takes place. Finally, the use of RPA for positioning in the current competitive environment is presented.

3.1 Development of the Competitive Environment

3.1.1 Consideration of the Interest Rate Perspective

The low interest rate policy over the negative prime rate and the bond purchases by the ECB have been of particular importance for the earnings situation of banks in recent years (Buhse, 2015). As long-term yields fell into the negative range and their implications for the overall interest rate landscape, a significant earnings component of the credit institutions was reduced: the result of maturity transformation and the resulting structural contribution. For this purpose, (formally) short-term customer funds or borrowings on the money and capital market are invested long-term (Grill et al., 2022). At times, 10-year construction financing was offered by banks at an effective interest rate of less than 1.0% (Barlage, 2022). It is almost self-evident that no pronounced interest income can be achieved with this.

Without these loans, the banks' and savings banks' cash holdings would have had to be parked in the ECB's accounts or in (newly built) vaults. The negative interest rates of the ECB and the foreseeable maintenance costs of new vaults made these options appear unattractive (ECB, 2021).

M. R. Smeets et al., *Robotic Process Automation in Use*,
https://doi.org/10.1007/978-3-658-45902-4_3

Thus, investing in (low) interest-bearing bonds and (real estate-secured) long-term loans was the lesser of two evils to economically survive the low interest rate phase. Nevertheless, the earning power of banks declined massively, which was reflected in the economic indicators of many institutions (German Federal Bank, 2021).

At the same time, however, it was also clear that this low interest rate phase would eventually end, as the ECB provoked a higher inflation rate with its monetary policy. The devaluation of money accelerated (at the latest) since the beginning of 2022 globally—including in the Eurozone. In addition to the ECB, other central banks had also implemented a very loose interest rate policy, which changed significantly during the course of 2022 and in some cases even earlier (Tageschau, 2022). The massive increases in interest rates—due to the reduction or stopping of bond purchases, the long-term area is also massively affected—lead to a dilemma for the banks' holdings: their bonds and customer loans, which are still running for years, lose value, measured against current deals. To put this development into perspective, one must realize that the interest rate development of the first half of 2022 was so pronounced that it exceeded even the assumed interest rate shocks of the supervisory authority in terms of the amount of the interest rate increase.

As far as these assets are part of the current assets, the strict lower of cost or market principle of § 253 IV HGB applies. If they are assigned to the fixed assets, the mitigated lower of cost or market principle (§ 253 III HGB) applies, so that there is a right to choose depreciation. However, if the book value exceeds the total present value of the bank book, a provision for impending loss must be made in the commercial and tax balance sheet, as far as the bank currently already controls on a present value basis. Smaller banks, whose binding present value control only starts in 2023 with the abolition of the so-called Annex regulation—according to BaFin letter of 03.12.2021—could still carry out the provision formation on a conventional basis in 2022. As far as risks occur in the P&L consideration that cannot be depicted by devaluations, however, there is already an obligation for small institutions to form provisions today (IDW, 2017; BaFin, 2021). If these assets remain in the portfolio until the end of their term and flow back as promised and planned, "only" opportunity costs arise during the term, as more attractive investment options were waived.

The temporary negative effects can indeed be prevented with the conclusion of hedging transactions, but these in turn lead to (permanent) P&L burdens. With the resurgence of interest rates, the long-term commitments jeopardize the earnings situation of banks (again) in the short term, as they can lead to massive value corrections. The fact that rising interest rates can strengthen the earning power of banks in the medium term remains unaffected by this (German Federal Bank, 2019a, b).

3.1.2 Regulatory Perspective

The requirements of Basel III pose an additional burden on the banking world. This set of rules has made a plethora of (extensive) regulations relevant for credit institutions. This includes the tightening of capital requirements to make the entire industry more crisis-resistant. Thus, the interest-bearing business of banks is (further) restricted (Deutsche Bundesbank, 2011). The logic of regulatory law does not necessarily follow economics. Article 114 of the CRR stipulates that claims against member states in national currency do not require capital backing, as they are defined as secure (EU, 2022). In light of the difficulties that, for example, Greece had during the last financial crisis and which could only be averted with massive aid packages (European Commission, 2022), this assessment seems inconsistent.

At the same time, the regulator introduced the Leverage Ratio, which disregards risk weighting and demands a ratio of core capital to total exposure to be maintained (Deutsche Bundesbank, n.d.). The logic is hard to understand when on one hand, complex procedures are required (or supposed to be) to determine risk weights, and these are overridden by another procedure (= ratio). One could almost get the impression that the regulator only attributes limited control effect to its introduced instruments. One could also pointedly ask why complex ratings are introduced if the results of risk-adjusted capital backing are not believed (Ostendorf, 2023). For small institutions, the implementation of the multitude of requirements—despite existing reliefs—is particularly difficult, as they only have (very) limited manpower to implement these issues (BVR, 2021, 2022). This apparent disadvantage is also economically hard to understand, as in the last financial crisis, the small units were not responsible for (system-relevant) dangers. This pressure to merge leads to a loss of granularity, making future crises more difficult to manage. Against this background, the political idea of merging Commerzbank with Deutsche Bank seems economically rather questionable (Beise, 2019; Gammelin, 2019). Therefore, it is good that this topic is no longer part of the current political agenda.

3.1.3 Moral Claim Dilemma

The divergence between desire and reality in the area of requirement fulfillment poses another challenge. In addition to economic advantages in an investment, compliance with other criteria is increasingly demanded. These include ecological, ethical, and social minimum standards. The growth rates of these investment products are remarkable, even if the overall level is still comparatively low. To meet this investor demand, indices have been developed, such as the *DAX® 50 ESG, the* MSCI® World SRI Index, and the Dow Jones Sustainability Index Worldwide (DJSI®) (Ostendorf et al., 2023). This increased societal demand is also reflected in other areas, such as agriculture and the production process with foreign suppliers (Zander et al., 2013; BMEL, 2015; Sepehr, 2018)

At the same time, there is a tendency of active law violations, which also became apparent. These include

- the VW diesel scandal, which of course does not affect the banking industry, but illustrates the extent of possible legal violations (NDR, 2020),
- the CumEx violations, where missing regulations were used for massive tax evasion (Jessen, 2019),
- extensive legal violations in major customer relationships (RND, 2022),
- investor deception during the 2008 financial crisis (RND, 2022),
- involvement in money laundering (RND, 2022),
- illegal rate setting of significant money market interest rates (RND, 2022),
- ignoring international trade bans (RND, 2022),
- the (alleged) greenwashing in the sustainable DWS funds (Berger, 2022).

As a result, these scandals have likely significantly damaged the image of the banking industry.

> **Important** In summary, it can be stated that the proclaimed claim has significantly increased in recent years; however, in lived reality, even the laws are often not complied with. In the context of this tension, it is necessary to align oneself. However, one's own ambitious orientation is endangered by the (mis-) behavior of competitors, as a certain transfer effect as collective punishment cannot be (completely) ruled out (Mussler, 2022).

3.1.4 Influence of Consumer Advocates

Against the backdrop of the sometimes (massive) legal violations and customer disadvantages by banks, it is hardly surprising that consumer advocates also took and continue to take a close look at banks. As understandable as the motivation and intention are, these initiatives sometimes lead to bizarre results. Today, for the sake of consumer protection, banks are no longer allowed to grant loans secured by real estate to persons over 60 years of age, unless their complete repayment can be guaranteed. To make it clear: A homeowner who wants to renovate his residential property—which is fully paid for and represents a first-class bank security—at the age of over 60, no longer gets a loan from a bank (Consumer Center, 2017). Pragmatic solutions are obvious, by having another, younger person, for example a child of the homeowner, act as a (co-)borrower.

Another workaround is to sell half of the property to a company specialized in this. Purely economically, this construct, which incurs land acquisition tax and other transaction costs through the sale, is an expensive variant for the (private) borrower. The financing specialist refinances himself through a classic mortgage-secured bank loan. Without going into details, it is self-explanatory that the "detour financing" must be less attractive for the end customer, as Fig. 3.1 visualizes (Dräbing, 2022; Sebastian et al., 2022).

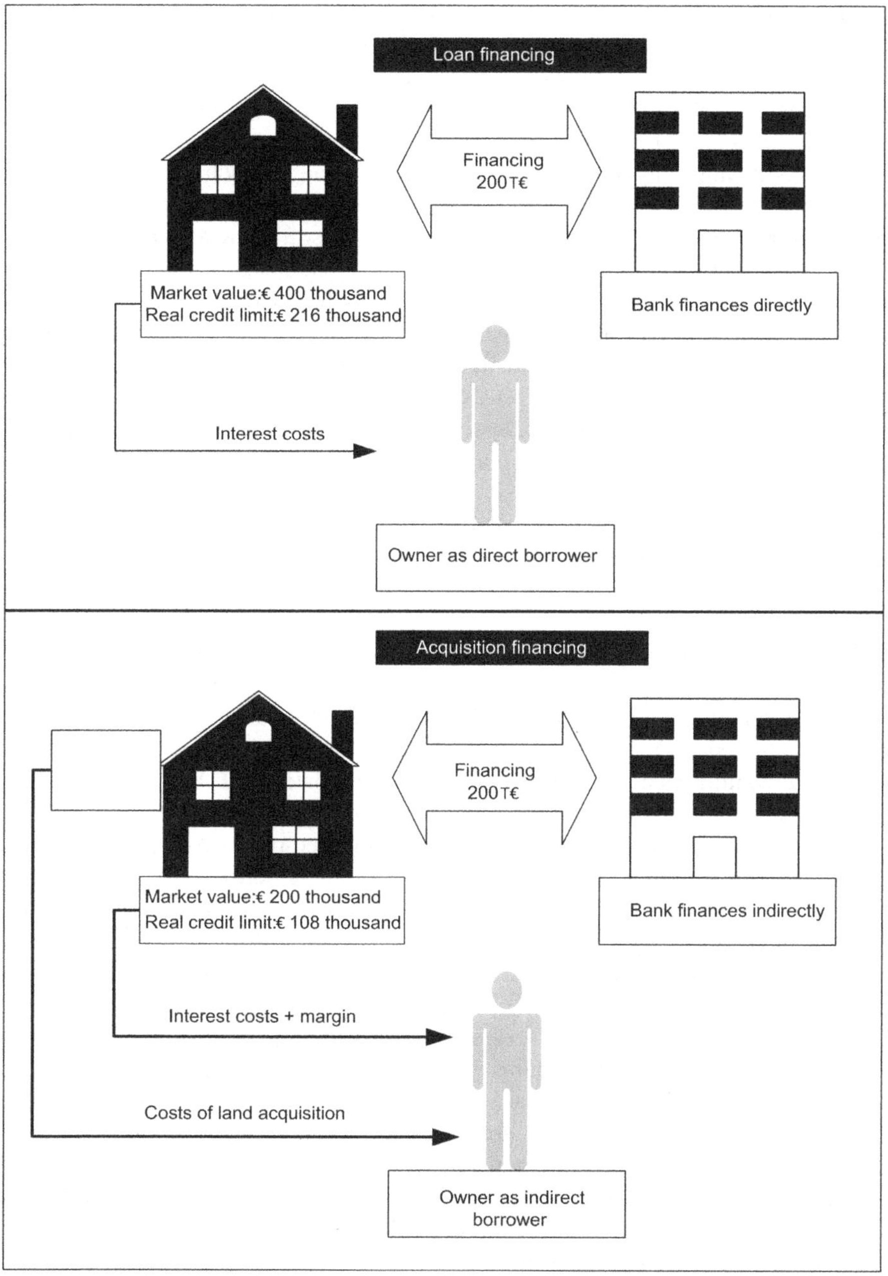

Fig. 3.1 Systematic comparison of direct to indirect real estate financing

Due to the system, the loan amount alone is significantly higher compared to the (half) lending object, so that the loan amount tends to exceed the real credit limit (Ostendorf, 2014; Grill et al., 2022). In addition, there is the margin claim of the interim financier as well as the costs of land acquisition incurred by him, which are also to be borne by the original borrower. In total, the (older) customer pays a significantly higher price for a detour financing, which could also be represented with a classic bank product and is also more cost-effective. At times, it seems as if consumer protection has become self-sufficient and it is no longer about protecting the end customer, but against the interests of the bank and customers. The customer has a financing need, which can be covered by a classic bank product, but which is no longer legally compliant. Since the need of many customers still exists, another and more expensive business model for the consumer has established itself, which was only questioned by consumer advocates in the summer of 2020 (Consumer Center, 2022b). And this despite the fact that, in addition to the named cost disadvantages—depending on the contract design—the appropriation of the value increases of half the property for the interim financier is also provided (Dräbing, 2022; Sebastian et al., 2022). If the banks fail to acquire these interim financiers as customers, they also lose a secure business.

3.1.5 Further Operational Complications

The BGH ruling of April 27, 2021 (AZ: XI ZR 26-20) has also imposed a higher administrative burden on banks by declaring the previously practiced general terms and conditions clauses as invalid. These stipulated that the customer had to object to a change within a given period; if they did not, the new regulation was considered accepted. As a result of the aforementioned ruling, every credit institution is now obliged to explicitly obtain the customer's consent (Federal Court of Justice, 2021). Ultimately, this is associated with two consequences: the account-holding banks must move an extreme wheel of administration to actually reach all customers and obtain their consent. At the same time, the flood of information with which the customer is confronted increases, as does their administrative effort. Whether this adjustment serves the customer and leads to greater transparency of the legal adjustment remains to be seen (LTO, 2022).

Many crowdfunding platforms are specifically designed so that they do not require BaFin's permission and are consequently not supervised by it. These service providers are therefore attributable to the grey capital market, which is characterized by a lack of state regulation. Here too, cost disadvantages arise for the established players. In addition, banks have stricter anti-money laundering requirements to meet than other companies, such as precious metal dealers. The same applies to online payment services and trading platforms, which also do not have to comply with/ have weakened banking supervisory regulations (Hirt & Kröner, 2019). These higher costs must also be shouldered and thus priced into the conditions.

The increasing digitization requires additional investments and thus implies write-offs of subsequent periods. At the same time, it endangers existing investments in existing bank infrastructure (= branches). The trend towards online banking was intensified by the Corona pandemic with the state-imposed restrictions (Wunsch-Weber & Zdrzalek, 2022).

Due to high inflation, both savings banks and cooperative banks see the saving ability of large sections of the population in Germany as endangered. The political demand of the Greens to legally regulate the current account interest rates, which currently average at 10% (should), is understandable given the discrepancy to the current interest rate level (end of 2022). On the other hand, credit institutions—as already explained here—are currently operating in an economically difficult environment. Ultimately, politicians must also be aware that banks need sources of income to finance the legal and regulatory requirements imposed on them (o.V., 2022).

3.2 Suitable Competitive Strategies for the Current Environment

Here the question arises as to the performance profile that the respective customer expects from their bank or savings bank. This is significant insofar as a divergence between the performance promise perceived by the customer and the actual service provided by the credit institution produces dissatisfaction (Stauss & Seidel, 2014).

3.2.1 Porter's Approach as a Starting Point

The starting point of the consideration are Porter's generic strategies, which have a significant influence on the discussion of competitive positioning. According to Porter's concept, a company must decide whether it

- operates in the overall market or a segment (= niche) and whether it
- offers its customers a favorable price—in connection with the basic benefit—or added value—in combination with an increased price.
- The basis for the favorable price are low costs, which are based, among other things, on experience curve gains. The corresponding strategy is called cost leader.
- There are various starting points for creating added value, such as quality, service, etc. It is important that the additional benefit generated from the customer's perception also leads to a higher willingness to pay. This type of strategy is called a differentiator (Porter, 2013, 2014).

In the following Fig. 3.2, the overall context is shown, in which the two dimensions of price and product benefit are plotted.

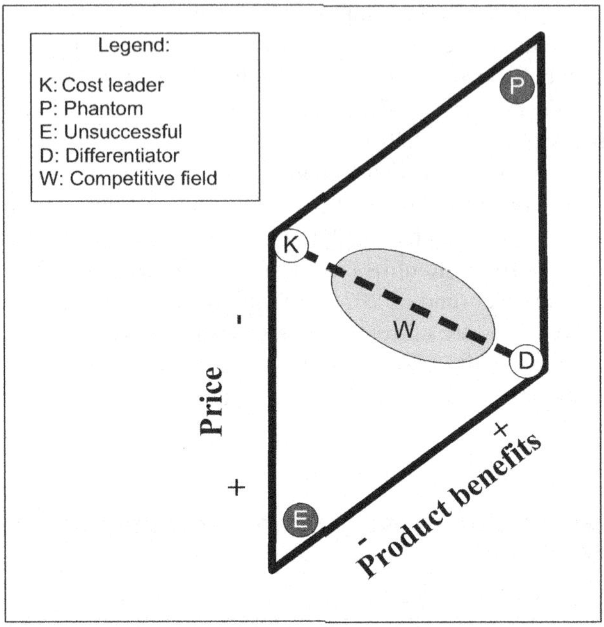

Fig. 3.2 Visualization of Porter's positioning options

The abbreviations stand for the name of the strategy type: thus, ***K*** symbolizes the cost leader and ***D*** the differentiator. The company that takes the position ***D*** offers the maximum additional benefit in combination with the highest price. In the banking world, the private bank with maximum individualized service and corresponding consulting performance is likely to be located here (NTV, 2022). The position ***K*** stands for the opposite combination of price and product benefit. The internet bank with minimal fees and equally pronounced service and consulting offer. Porter attests to all other orientations no successful positioning in the competition, which is also evident in the generic claim of his strategies (Raabe, 2008; Porter, 2013, 2014).

In Fig. 3.2, the points ***K*** and ***D*** are connected by a dashed line: this represents the positioning line depending on the combination of price and benefit. Of course, there can always only be one (absolute) cost leader. A company that combines a slightly higher price with a slightly higher benefit undoubtedly does not realize the extreme position, but should still be classified as a cost leader. The internet bank with (slightly) increased service may even be more successful than if the offer is too sparse. If the focus on costs is exaggerated, there is a risk of falling short of the required minimum standard. The same risk exists for the differentiator at the moment when he overtaxes the willingness of his (potential) customers to pay. In the perception of the authors, many banks were traditionally positioned in the field ***W*** (competition) along the competition line. A combination of high price and low product benefit—symbolized in Fig. 3.2 by ***E*** (= unsuccessful)—has no real charm in competition and therefore does not require further consideration as a desirable positioning. The opposite point, which is unattainable in Porter's reading and is therefore referred to as ***P*** (phantom) in Fig. 3.2, is different (Porter, 2013, 2014; Ostendorf et al., 2021).

3.2.2 Hybrid Strategies

This assessment that a low price in combination with a high benefit is incompatible and therefore utopian, as well as the generic claim, have provoked various contradictions, which also led to the development of new approaches (Fleck, 1995; Mays, 2018).

One form is the multilocal approach, which uses the two Porterian strategy types in pure form. Depending on the spatial conditions, however, both forms are in use, as Fig. 3.3 illustrates (early on: Perlmutter, 1969; Meffert, 1986).

A regionally oriented credit institution—for example, a cooperative bank or a savings bank—acts as a differentiator (***D***) in its own business area by offering a high standard of consulting quality and service. Outside of its traditional business area, it operates—within the scope of its legal possibilities—as a cost-oriented institution (***K***), by foregoing branches there and acting like an internet bank. In this context, the reference to NRW and the spatial positioning of the differentiator is only to be seen as an example and thus does not represent a reflection of real competitive conditions, although this approach will certainly occur in practice. This approach has a certain charm; the original business field is worked on with the previous strategy and the other customers receive a light product, which however

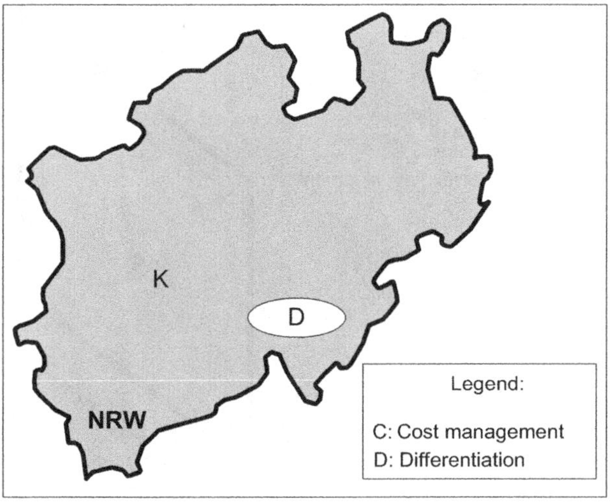

Fig. 3.3 Exemplary multilocal positioning

helps the credit institution to realize size advantages in the use of the customer service center, the market sequence etc.

Sequential hybrid strategies are those that, among other things, go back to, e.g., Gilbert and Strebel and are characterized by the fact that the two advantage dimensions of price and product benefit are focused on one after the other. If the cost leader (***K***) in Fig. 3.1 can no longer exploit any further (significant) cost reduction potentials (temporarily), there is the possibility to increase the product added value. If it succeeds in keeping the cost level low, there is a parallel development towards the phantom (***P***) (Gilbert & Strebel, 1987; Kleinaltenkamp, 1987). To what extent this development can be continued probably depends on the individual case. A classic internet bank can certainly improve its range of services, but will reach limits.

The same approach can be pursued from the perspective of the differentiator, by realizing cost improvements at a given (high) benefit level. The differentiator (***D***) rises vertically in the direction of the phantom. Here too, the degree of realization—especially in the service sector—must remain open. The structures of a private bank equipped with branches and managing assets prevent the copying of the cost structures that are inherent to an internet bank. An overview is provided by Fig. 3.4.

The figure clearly shows that in addition to the companies that take extreme positions, the companies from the competitive field (***W***) can also move towards the phantom, which realize cost reductions and/or benefit increases.

For the industrial sector, the so-called simultaneous hybrid strategies were created, which aim to achieve higher flexibility than previously usual through intelligent manufacturing. This should generate both low costs and a high original

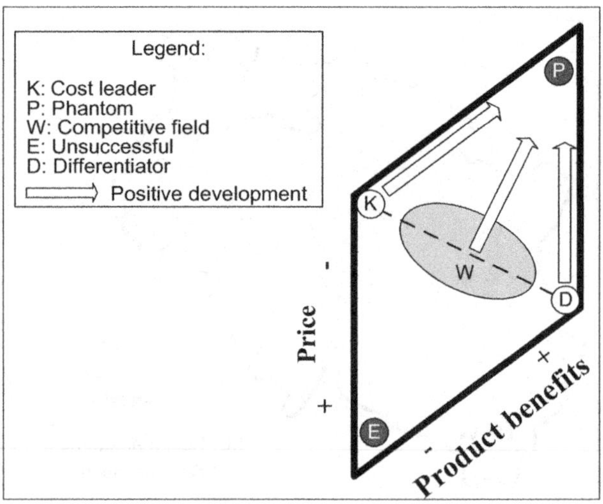

Fig. 3.4 Sequential hybrid development

product benefit (Kaluza, 1989; Corsten & Will, 1994). This balancing act between the two dimensions of price and product benefit is difficult to achieve for the banking sector, as the branch network is a significant cost driver (Wunsch-Weber & Zdrzalek, 2022).

If you compare these theoretical possibilities with reality, you will find that the banks were and are under considerable economic pressure. In order to still be successful, there were and are some adjustments. Some of these measures are:

- Introduction of negative interest rates for both corporate and private customers when certain thresholds are exceeded (Consumer Advice Centre, 2022a). To avoid any misunderstanding: this measure is economically correct from the perspective of the credit institution. One of the basics of business administration is that an offer that does not even cover variable costs should not be expanded (Olfert, 2018; Haberstock, 2022). From the customer's perspective, who had to accept not only the penalty interest but also the devaluation of money, this condition design is a burden (Consumer Advice Centre, 2022a). Some actors have positioned themselves more cleverly and sought interest-preserving solutions with the customer by offering profitable investment opportunities despite the low-interest phase (Wunsch-Weber & Zdrzalek, 2022). To what extent such solutions are generally suitable is a further question.
- The issue of branch closures also regularly leads to discussions. On the one hand, their maintenance is expensive, and on the other hand, the trend towards online banking has intensified. Which factor is the "chicken" or "egg" here can certainly be discussed. In times of crisis, cost-cutting is often the method of choice, leading to various branch closures. At the same time, however, the loss of a distinguishing feature also goes hand in hand with branch closures. Whether the customer perceives this feature and is especially willing to pay a higher price for it probably depends on the individual case. The Volksbank Frankfurt am Main and the Sparkasse are taking a different approach by using the same branches on different days with individual lighting and staffing (Wunsch-Weber & Zdrzalek, 2022)
- From the authors' own experience, it is known that at the beginning of the century the claim for telephone banking was 80 ⇔ 20: This means that 80% of incoming calls are to be answered within 20 seconds. Ambitious banks even aimed for the 90 ⇔ 10 rule (Callcenter Boerse GMBH, n.d.). In the current present, waiting times of more than 10 minutes are not uncommon and are justified with the note that one would also have to wait more than 60 minutes at the mobile phone provider's hotline.
- To reduce process costs, securities were massively waived; even customers classified the banks' lending as generous (Ostendorf et al., 2018).
- In some account management models, the number of free monthly cash withdrawals is limited (Atzler, 2019). Supermarkets use this measure to offer the possibility of taking cash (in limited amounts) with card payments (Leistikow, 2020). Here, some banks have given up parts of the value chain and customer contact, and the entire industry now has to deal with the changed competition.

- The fee rates for confirmed LZB checks, which serve, among other things, for participation in public auctions, are sometimes 45,- €; and this even though the customer is debited the equivalent value before the issue and he has to liquidate the unused check via the normal (fee-based) submission process (Ostendorf, 2014; Grill et al., 2022; Sparkasse Fürth, 2022).

If one summarizes the tendency of this development, some credit institutions have given up their previous positioning and have developed in the direction of "unsuccessful". This tendency is shown in Fig. 3.5.

With the outlined development—if it is correct—there is no criticism of the acting persons. The authors are aware that the environmental conditions were and are extremely difficult. In addition, achieving an appropriate operating result at the operational level often has the highest priority.

3.2.3 Extended Consideration

Another strand of strategy development is the inclusion of ecology. Many of these approaches date back to the 1990s and 2000s. In some cases, ecology was included as a third dimension in the competitive consideration, as it fundamentally differs from the original product benefit: the purchaser of an ecological or more ecological product does not have an individualized right of use. In other words: he cannot exclude other consumers from the advantage he has created (Hummel, 1997; Ostendorf, 2000; Mays, 2018).

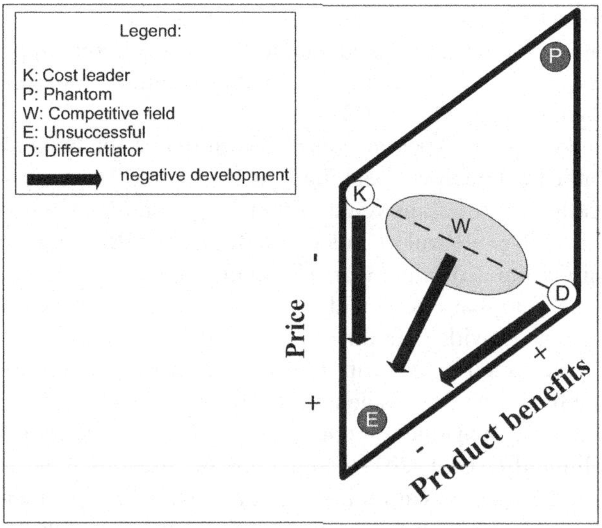

Fig. 3.5 Erosion of competitive position over time

Since the starting points for (significant) ecological improvements in the banks' performance creation process are less pronounced than in production or logistics companies, the relevance of these strategies for the banking sector was clear (Ostendorf, 2000). Instead of making their own business operations ecological, banks—through ecological requirements in lending indirectly via their customers—have a much larger lever available.

The banking supervision has now gone so far as to want sustainability, and in particular climate risks, to be integrated into risk-bearing capacity. To this end, the EBA published guidelines in 2020, which demand the principal consideration of ESG risks (EBA, 2020).

There are now some so-called eco-banks that use ethics and ecology as loan granting criteria and also align their production processes accordingly (GLS, n.d.; Utopia, n.d.). Traditionally oriented banks must also engage in the financing of ecologically oriented loans. However, (potential) savers have no certainty with these competitors as to what their deposits are really financing.

This has called the EU into action: Under the label of consumer protection, Regulation (EU) 2021/1253 was introduced, which is legally effective from 02.08.2022 and demands the inquiry of sustainability preferences in the context of consultation (EU, 2021). This involves a fundamental intervention in the consultation process, which is difficult to reconcile with a liberal worldview (Eisenring, 2022). Moreover, the concept of sustainability is very dazzling. Thus, as of July 2022, membership in the nuclear industry is (still) an exclusion criterion for listing in the DAX® 50 SEG—the sustainable member of the DAX® family (Börse Frankfurt, 2022; Deutsche Börse, n.d.). However, on 06.07.2022, the EU Parliament also classified investments in natural gas and nuclear energy as sustainable. There is massive resistance against this. For example, Greenpeace wants to sue against the classification, as it is not legally compliant and would mislead consumers (Greenpeace, 2022). Further regulations are being prepared by the EU: In the future, requirements are to be defined both in terms of social issues and corporate governance. It is to be feared that this will create the next bureaucratic hydra (Eisenring, 2022). Elon Musk has an even more drastic assessment on Twitter, as Capital reports (Laube, 2022).

> **Important** What consequences does this now have for the banks? The requirements for ecology, social claim, and corporate governance can be subsumed under societal claim. It goes without saying that compliance with laws represents the relative zero point. However, the explanations at the beginning of the chapter have made it clear that companies can also align themselves below this level if the (supposed) economic incentive is large enough. For the corporate orientation, it follows that the previously presented graphic on corporate orientation needs to be expanded by one dimension, as Fig. 3.6 illustrates.

The quadrant from ***K***, ***D***, ***P*** and ***E*** of the previous representation is to be aligned along another dimension, that of societal claim. The illustration makes it clear that one of the two visualized quadrants implements the relative zero point of mere

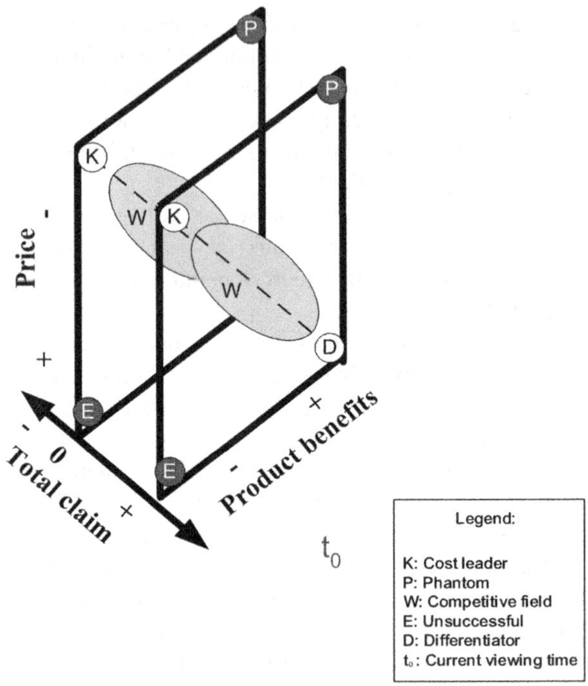

Fig. 3.6 Three-dimensional competitive field

legal compliance, but does not go beyond it. The second quadrant of the classic competitive environment realizes a higher societal benefit, even if its actors have not yet reached the maximum. The individual credit institution must find its positioning also in the dimension of societal claim level—possibly depending on its group affiliation and the associated potential collective guilt in public perception. The dashed circle around the entire positioning field represents the current status quo of societal demands and knowledge and is marked as t_0.

Improvements are now possible in the direction of the three dimensions, as Fig. 3.7 illustrates.

Thus, an improvement in the cost situation, the perceived quality, and the fulfillment of societal expectations can occur. This is illustrated by the arrows in the figure. The gray area indicates that any point in the three-dimensional coordinate system is achievable, as long as it lies on the surface or in the space below. RPA is predestined for such improvements, as the explanations in Chap. 2 have already shown. If personnel costs are substituted by significantly lower material costs, the total costs of the process can decrease. At the same time, a smaller error rate leads to an increase in quality from the customer's perspective. As far as the freed-up employee capacities take on more demanding tasks, this measure strengthens the social profile of the bank. At the same time, a more reliable process leads to higher legal certainty with regard to meeting the requirements of the supervisory

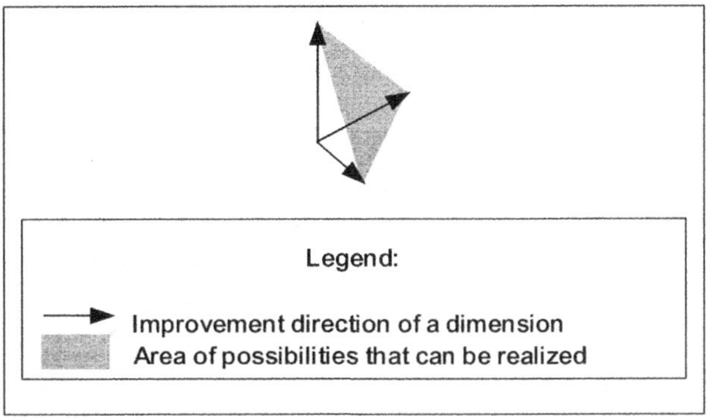

Fig. 3.7 Possibilities for improvement in three-dimensional space

authority (Chap. 4). Strategically, this can stop the process of outsourcing and possibly even reverse it, by increasing the value contribution of the credit institution (again).

Examples of such processes are:

- The automated creation of new accounts with RPA, which were previously applied for online and where online legitimation of the customer has taken place
- The faster processing of unstructured customer inquiries—with the help of RPA and an AI component

Whether the management of the individual institutes takes these opportunities or uses these starting points for further outsourcing remains to be seen.

> **Important** The special charm of RPA is that it can provide added value at any strategic positioning level. With this, a company from the ***Unsuccessful*** area, a (pure) ***Cost leader*** as well as a (pure) ***Differentiator*** can further profile themselves, just like companies that are positioned in the *Competitive field.*

Thus, an adequate tool is available to meet the current dynamics. While the dynamics of the 1990s were still characterized by new competitors and products (Kaluza, 1996; Mays, 2018), today it is necessary to deal appropriately with structural breaks. In addition to the Ukraine war and the Corona pandemic, whose effects were already topics, the EU is a legislator at work, which sometimes formulates very bureaucratic requirements. In addition, the present is characterized by a—in its speed—previously unknown increase in knowledge (o.V., 2019), so that companies must always act flexibly to adapt to the changed environment of knowledge and demands. The transformation process from the present $\mathbf{t_0}$ to a future point $\mathbf{t_n}$ is visualized in Fig. 3.8. The starting point is the company's positioning in the present against the background of the current conditions in the

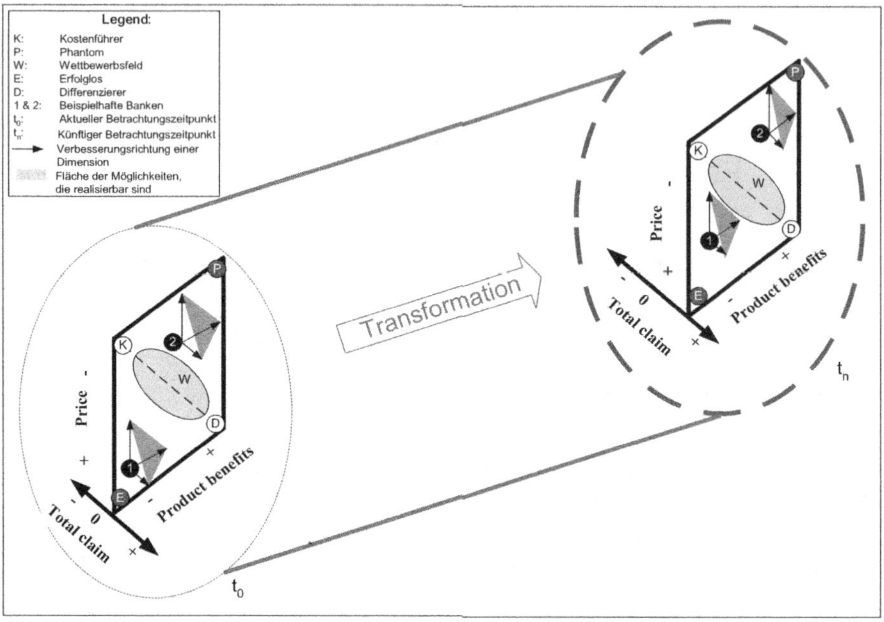

Fig. 3.8 Transformation process from t_0 to t_n

competition quadrant of costs and differentiation. In the status quo, the players considered are slightly more ambitious than the legal standard in terms of the third dimension. This is evident from the positioning of the quadrant beyond the zero point of the societal expectation level. Company 1 has a less attractive competitive position and tends more towards the point Unsuccessful. Company 2 is much better positioned and offers a (high) added value at comparatively low costs. Both improve their position with the help of RPA, resulting in a transformation to the point t_n.

At the point t_n, the framework conditions have changed and the companies considered are now operating in a quadrant that generates a significantly higher societal benefit than at the starting point. At the same time, both companies were able to position themselves better in the tension field of costs and additional benefits, so that they can record a significant improvement in their competitive position despite more demanding environmental conditions.

3.3 Summary

The aim of the chapter was to clarify the current framework in which credit institutions must position themselves. The result is that the industry has come under increasing pressure and must implement requirements initiated by other players. These contracts at the expense of third parties have endangered and continue to endanger the economic success of the banks. Adjustments in the industry, which rarely benefited the customer, were made out of this necessity. Whether the measures applied on a case-by-case basis are generally suitable cannot be assessed.

In total, the competitive situation of many industry participants has significantly deteriorated, as the profile for which they stand has eroded. As early as 2019, every second customer perceived his bank or savings bank as interchangeable (Leichsering, 2019). Even if no new trend towards individualization of banking business is to be expected from the use of RPA, the technology can contribute to sharpening the profile and thus positioning.

The developments shown were compared with common competitive strategies and their consequences determined. In terms of content, it was shown that a three-dimensional orientation in competition currently appears promising. This approach was visualized and a possibility of transformation from the present to the future was demonstrated.

- The competitive situation for banks is intensifying; many former competitive advantages of established banks are no longer (to the same extent as a few years ago) present
- This creates pressure to examine and possibly make changes in strategic positioning
- Possible approaches can be examined and discussed using the well-known Porter's model of competitive strategies
- RPA offers the opportunity to streamline processes, reduce costs and possibly bring new products to market quickly
- In this way, RPA can create competitive advantages and thus contribute to the strategic (re-) positioning of the institutes

References

Atzler, E. (2019). *Immer mehr Banken verlangen Geld für Barabhebungen*. https://www.handelsblatt.com/finanzen/banken-versicherungen/banken/bankgebuehren-immer-mehr-banken-verlangen-geld-fuer-barabhebungen/25274156.html. Accessed: 16. Aug. 2022.

BaFin. (2021). Aufsichtliche Beurteilung interner Risikotragfähigkeitsverfahren, GZ: BA 54-FR 2210-2021/0007.

Barlage, B. (2022). *Zinsentwicklungen und Expertenmeinungen: Hier finden Sie wichtige Infos & aktuelle Bauzinsen*. https://www.interhyp.de/ratgeber/was-muss-ich-wissen/zinsen/zinscharts/. Accessed: 3. Sept. 2022.

Beise, M. (2019). *Bankenfusion—Es ist eine Schande, was Scholz da macht*. https://www.sueddeutsche.de/wirtschaft/scholz-deutsche-bank-commerzbank-fusion-1.4374177. Accessed: 3. Sept. 2022.

Berger, D. (2022). *„Greenwashing-Affäre". Grüne Täuschung bei der DWS-Gruppe?*https://www.zdf.de/nachrichten/wirtschaft/greenwashing-deutsche-bank-woehrmann-100.html. *Accessed: 4. Sept. 2022.*

Börse Frankfurt. (2022). *Nachhaltig investieren*. https://www.boerse-frankfurt.de/nachhaltigkeit. Accessed: 7. July 2022.

Buhse, M. (2015). *Die Schwachstellen in Draghis Milliardenplan*. https://www.zeit.de/wirtschaft/2015-03/kauf-staatsanleihen-europaeische-zentralbank. Accessed: 5. July 2022.

Bundesgerichtshof. (2021). Urteil AZ XI ZR 26-20 vom 27. April 2021.

Bundesministerium für Ernährung und Landwirtschaft—Gutachten des Wissenschaftlichen Beirats für Agrarpolitik beim Bundesministerium für Ernährung und Landwirtschaft. (2015). *Wege zu einer gesellschaftlich akzeptierten Nutztierhaltung*. https://www.bmel.de/SharedDocs/Downloads/DE/_Ministerium/Beiraete/agrarpolitik/GutachtenNutztierhaltung.pdf?__blob=publicationFile&v=2. Accessed: 3. Sept. 2022.

Bundesverband der Deutschen Volksbanken und Raiffeisenbanken. (2021). *Vorschlag zu Basel III mit Licht und Schatten*.

Bundesverband der Deutschen Volksbanken und Raiffeisenbanken. (2022). *BVR kritisiert den Berichtsentwurf zur Umsetzung von Basel III: Gegebenheiten des europäischen Bankenmarktes werden zu wenig berücksichtigt*.

Callcenter Boerse GmbH. (n.d.). *Tipps zum Outsourcen an ein Call-Center*. https://www.call-center.de/hotline-infoline-marketing-tipp-34.html. Accessed: 16. Aug. 2022.

Corsten, H., & Will, T. (1994). Simultaneität von Kostenführerschaft und Differenzierung durch neuere Produktionskonzepte—Informationstechnologisches und arbeitsorganisatorisches Unterstützungspotential. *zfo, 63*, 286–293.

Deutsche Börse. (n.d.). *Neuer DAX setzt auf Nachhaltigkeit*. https://www.deutsche-boerse.com/dbg-de/me-dia/pressemitteilungen/Neuer-DAX-setzt-auf-Nachhaltig-keit-1786626. Accessed: 7. July 2022.

Deutsche Bundesbank. (2011). *Basel III—Leitfaden zu den neuen Eigenkapital- und Liquiditätsregeln für Banken*.

Deutsche Bundesbank. (2019a). *Ergebnisse des LSI-Stresstests 2019—Gemeinsame Pressenotiz mit der BaFin*. https://www.bundesbank.de/de/presse/pressenotizen/ergebnisse-des-lsi-stresst-ests-2019-807574. Accessed: 3. Sept. 2022.

Deutsche Bundesbank. (2019b). *Ergebnisse des LSI-Stresstests 2019—Präsentation zum gemeinsamen Pressegespräch*. https://www.bundesbank.de/resource/blob/807590/9c2dd30ac74d686ec4a741117b763166/mL/2019-09-23-stresstest-anlage-data.pdf. Accessed: 10. Sept. 2022.

Deutsche Bundesbank. (2021). *Die Ertragslage der deutschen Kreditinstitute im Jahr 2020*. https://www.bundesbank.de/resource/blob/876222/0c1a444b427cf75d90deb2f6df794808/mL/2021-09-ertragslage-data.pdf. Accessed: 3. Sept. 2022.

Deutsche Bundesbank. (n.d.). *Leverage ratio*. https://www.bundesbank.de/de/aufgaben/bank-enaufsicht/einzelaspekte/leverage-ratio/leverage-ratio-598484. Accessed: 6. July 2022.

Dräbing, T. (2022, April 19). Teuer und riskant. *Berliner Zeitung*.

Eisenring, C. (2022). *Atomkraft und Gas gelten in der EU künftig als nachhaltig. Doch die Übung bleibt ein bürokratischer Albtraum*. https://www.nzz.ch/meinung/taxonomie-mit-gas-und-atomkraft-eu-betreibt-schadensbegrenzung-ld.1692369. Accessed: 7. July 2022.

Europäische Kommission. (2022). *Enhanced surveillance framework for Greece*. https://ec.europa.eu/info/business-economy-euro/economic-and-fiscal-policy-coordination/financial-assistance-eu/which-eu-countries-have-received-assistance/financial-assistance-greece_de. Accessed: 6. July 2022.

Europäische Union. (2021). *Delegierte Verordnung (EU) 2021/1253 der Kommission vom 21. April 2021*.

Europäische Union. (2022). *Verordnung (EU) Nr. 575/2013 des Europäischen Parlaments und des Rates vom 26. Juni 2013*.

European Banking Authority. (2020). *EBA report on management and supervision of ESG risks for credit institutions and investment firms*.

EZB. (2021). *Der Negativzinssatz der EZB*. https://www.ecb.europa.eu/ecb/educational/explainers/tell-me-more/html/why-negative-interest-rate.de.html. Accessed: 5. July 2022.

Fleck, A. (1995). *Hybride Wettbewerbsstrategie. Zur Synthese von Kosten—und Differenzierungsvorteilen*. Deutscher Universitätsverlag.

Gammelin, C. (2019). *Warum Olaf Scholz die Gründung einer Superbank gut fände*. https://www.sueddeutsche.de/wirtschaft/deutsche-bank-commerzbank-fusion-1.4372779. Accessed: 3. Sept. 2022.

Gilbert, X., & Strebel, P. (1987). Strategies to outpace them competition. *TJoBS, 9*, 28–36.

GLS. (n.d.). *Die Geschichte der GLS Bank*. https://web.archive.org/web/20120104210058/http://www.gls.de/die-gls-bank/ueber-uns/geschichte.html#c11370. Accessed: 12. June 2022.

Greenpeace. (2022). https://www.greenpeace.de/klimaschutz/klimakrise/eu-taxonomie-klage c11370. Accessed: 20. Sept. 2022.

Grill, W., Perczynski, H., Int-Veen, T., Menz, H., & Pastor, D. (2022). *Wirtschaftslehre des Kreditwesens*. Westermanngruppe.

Haberstock, L. (2022). *Kostenrechnung I*. Schmidt.

Hirt, O., & Kröner, A. (2019). *Internet-Konkurrenz setzt Banken zu*. https://www.fr.de/wirtschaft/internet-konkurrenz-setzt-banken-11258448.html. Accessed: 4. Sept. 2022.

Hummel, J. (1997). *Strategisches Öko-Controlling—Konzeption und Umsetzung in der textilen Kette*. Deutscher UniversitätsVerlag.

Institut der Wirtschaftsprüfer in Deutschland e. V. (2017). RS BFA 3 (n.F.).

Jessen, J. (2019). *Cum-Ex-Skandal. Viel mehr als ein Schlupfloch*. https://www.zeit.de/2019/51/cum-ex-skandal-steuerrecht-staatsversagen?utm_referrer=https%3A%2F%2Fwww.google.com%2F. Accessed: 4. Sept. 2022.

Kaluza, B. (1989). *Erzeugniswechsel als unternehmenspolitische Aufgabe. Integrative Lösungen aus betriebswirtschaftlicher und ingenieurwissenschaftlicher Sicht*. Schmidt.

Kaluza, B. (1996). *Dynamische Produktdifferenzierungsstrategie und moderne Produktionssysteme*. In Diskussionsbeitrag Nr. 211 des Fachbereichs Wirtschaftswissenschaft der Gerhard-Mercator-Universität-GH-Duisburg.

Kleinaltenkamp, M. (1987). Die Dynamisierung strategischer Marketing-Konzepte—Eine kritische Würdigung des „Outpacing-Strategies"-Ansatzes von Gilbert und Strebel. *zfbf, 39*, 31–52.

Laube, H. (2022). Der bewegte Mann. *Capital, 7*, 76–81.

Legal Tribune Online. (2022). *Ein Jahr nach Bankgebühren-Urteil des BGH. Hunderte Verbraucher beteiligen sich an Musterprozess*. https://www.lto.de/recht/nachrichten/n/bgh-urteil-fingierte-agb-banken-kontogebhren-ein-jahr-spter-klagen-beschwerden-verbraucher/. Accessed: 4. Sept. 2022.

Leichsering, H. (2019). *Für die Hälfte der Kunden sind die Finanzinstitute austauschbar*. In der Bank Block vom 12.02.2019. https://www.der-bank-blog.de/fuer-haelfte-kunden/strategie/37652573/. Accessed: 11. Sept. 2022.

Leistikow, D. (2020). *Weniger Geldautomaten: So bekommen Sie trotzdem Bares*. https://www.computerbild.de/artikel/cb-News-Finanzen-Geldabheben-Rewe-Aldi-Penny-31969887.html. Accessed: 16. Aug. 2022.

Mays, V. (2018). *Wettbewerbsstrategien. Eine vergleichende Analyse zwischen der Dynamischen Ökologieführerschaft und der Blue Ocean Strategie*. LIT.

Meffert, H. (1986). Marketing im Spannungsfeld von weltweitem Wettbewerb und nationalen Bedürfnissen. *ZfB, 56*(8), 689–712.

Mussler, H. (2022). *Rufschaden für die Sparkassen*. https://www.faz.net/aktuell/finanzen/steuerfahnder-bei-der-deka-rufschaden-fuer-die-sparkassen-18121071.html. Accessed: 8. July 2022.

NDR. (2020). *Die VW-Abgas-Affäre: Eine Chronologie*. https://www.ndr.de/nachrichten/niedersachsen/braunschweig_harz_goettingen/Die-VW-Abgas-Affaere-eine-Chronologie,volkswagen892.html. Accessed: 4. Sept. 2022.

NTV. (2022). *Der Börsentag. UBS enttäuscht mit Rekordgewinn*. https://www.n-tv.de/wirtschaft/der_boersen_tag/UBS-enttaeuscht-mit-Rekordgewinn-article23487766.html. Accessed: 26.07.2022.

O.V. (2019). *Explosion des Wissens: der Countdown läuft*. In KOM vom 18.03.2019. https://www.kom.de/medien/explosion-des-wissens-der-countdown-laeuft/. Accessed: 22.08.2022.

o.V. (2022). *Der Wohlstand bröckelt—Deutsche geraten an finanzielle Grenzen*. In Focus 22.08.2022. https://www.focus.de/finanzen/news/prognose-des-sparkassen-praesidenten-mehrheit-der-deutschen-wird-nicht-mehr-sparen-koennen_id_136872534.html?utm_source=newsletter&utm_medium=email&utm_campaign=newsletter_FINANZEN. Accessed: 22. Aug. 2022.

Olfert, K. (2018). *Kostenrechnung*. Kiehl.

Ostendorf, R. J. (2000). *Dynamische Ökologieführerschaft: eine Wettbewerbsstrategie gewinnorientierte Unternehmen—theoretische Darstellung und praktische Überprüfung am Beispiel der Automobilindustrie*. Verlag Wissenschaft und Praxis.

Ostendorf, R. J. (2014). *Bankwirtschaft—Grundlagen für Ausbildung, Praxis und Studium*. Pearson.

Ostendorf, R. J. (2023). Kommentierung Artikel 92a der CRR—Eigenmittelanforderungen und berücksichtigungsfähige Verbindlichkeiten für G-SRI. In R. Fischer & H. Schulte-Mattler (Hrsg.), *KWG CRR-VO—Kreditwesengesetz—VO (EU) Nr. 575/2013 Kommentar*.

Ostendorf, R. J., Rösen, A., & Rettich, L. M. (2018). *Kreditvergabepraxis der Sparkassen und Genossenschaftsbanken*. LIT.

Ostendorf, R. J., Smeets, M., Liepold, C., Bruns, E., & Rösner, M. (2021). RPA-Prozessverbesserungen: Taktische Position im Wettbewerb stärken. In R. J. Ostendorf (Hrsg.), *Nachhaltigkeit—differenzierte Perspektiven auf ein aktuelles Thema* (pp. 1–70). LIT.

Ostendorf, R. J., Liepold, C., & Schlöter, K. (2023). Nachhaltigkeit an der Börse. In R. J. Ostendorf (Hrsg.), *Krisenmanagement: Prävention, Identifizierung und Steuerung*. LIT.

Perlmutter, H. (1969). The Tortuous Evolution of the Multinational Corporation. *TCJoWB, 4*, 9–18.

Porter, M. E. (2013). *Wettbewerbsstrategie (Competitive strategy), Methoden zur Analyse von Branchen und Konkurrenten*. Campus.

Porter, M. E. (2014). *Wettbewerbsvorteile Spitzenleistungen erreichen und behaupten (Competitive Advantage)*. Campus.

Raabe, M. (2008). *Innovatives Bankmarketing—Erfolgsstrategien im Direct Banking*. Springer.

Redaktions Netzwerk Deutschland GmbH. (2022). *Epstein, Schwarzgeld, Zinsbetrug: Die größten Skandale der Deutschen Bank*. https://www.rnd.de/wirtschaft/deutsche-bank-die-grossten-skandale-HB2SV474SBCWTMU3XVDJ454QOM.html. Accessed: 6. July 2022.

Sebastian, S., Kirchhoff, M., & Herz, C. (2022, März 23). Taugt die Immobilie als Altersvorsorge? *Handelsblatt*, p. 32.

Sepehr, J. (2018). *Gerechtigkeit fordern. Made in Bangladesch: Fünf Dinge, die sich in der Bekleidungsindustrie ändern müssen*. https://www.globalcitizen.org/de/content/made-in-bangladesch/. Accessed: 3. Sept. 2022.

Sparkasse Fürth. (2022). *Preis und Leistungsverzeichnis vom 01.08.2022*. https://www.sparkasse-fuerth.de/content/dam/myif/spk-fuerth/work/dokumente/pdf/preise-leistungen/preis-leistungsverzeichnis.pdf?n=true. Accessed: 16. Aug. 2022.

Stauss, B., & Seidel, W. (2014). *Beschwerdemanagement*. Hanser.

Tagesschau. (2022). *EZB leitet Zinswende im Euroraum ein*. https://www.tagesschau.de/wirtschaft/ezb-leitzins-141.html. Accessed: 5. July 2022.

Utopia. (n.d.). *Die besten Ökobanken*. https://utopia.de/bestenlisten/die-besten-gruenen-banken/. Accessed: 12. June 2022.

Verbraucherzentrale. (2022a). *Verwahrentgelte—was Sie jetzt wissen müssen*. https://www.verbraucherzentrale.de/wissen/geld-versicherungen/sparen-und-anlegen/verwahrentgelte-was-sie-jetzt-wissen-muessen-64023. Accessed: 16. Aug. 2022.

Verbraucherzentrale. (2022b). *Teilverkauf der eigenen Immobilie: Was sind die Haken der Angebote?*https://www.verbraucherzentrale.de/wissen/geld-versicherungen/kredit-schulden-insolvenz/teilverkauf-der-eigenen-immobilie-was-sind-die-haken-der-angebote-48054. *Accessed: 16. Aug. 2022.*

Verbraucherzentrale Bremen. (2017). *Zu alt für einen Immobilienkredit?*https://www.verbraucherzentrale-bremen.de/pressemeldungen/zu-alt-fuer-einen-immobilienkredit-10066. *Accessed: 4. Sept. 2022.*

Wunsch-Weber, E., & Zdrzalek, L. (2022). *Diese Bankchefin hält nichts von Minuszinsen für Normalsparer*. https://www.wiwo.de/unternehmen/banken/frankfurter-volksbank-diese-bankchefin-haelt-nichts-von-minuszinsen-fuer-normalsparer/27973852.html. Accessed: 16. Aug. 2022.

Zander, K., Bürgelt, D., Christoph-Schulz, I., Salamon, P., Weible, D., & Isermeyer, F. (2013). *Erwartungen der Gesellschaft an die Landwirtschaft*. https://literatur.thuenen.de/digbib_extern/dn052711.pdf. Accessed: 3. Sept. 2022.

4 RPA from the Perspective of the Audit

After the use of RPA was classified in the third chapter in the context of the current challenges in the competitive environment of the banking and financial services industry, the fourth chapter focuses on the use of RPA in the specific context of audit requirements and the associated (supervisory) legal framework conditions. After a basic classification of RPA in terms of its already noticeable spread and relevance in the banking sector, a systematic classification of RPA as an audit object on the one hand and as a possible instrument of audit support on the other follows. For this purpose, the legal and regulatory requirements for the preparation of use and the specific RPA use are analyzed in detail in this chapter.

> **Important** This analysis should also precede the specific decision to use it, as it is often recognized too late in the euphoria of initial implementation successes of RPA that the regulatory requirements were either not or only insufficiently taken into account. Especially in the heavily regulated sectors of the financial industry, this weakness results in potential risks that should be avoided as much as possible (PWC, 2021).

The aim is therefore to provide both the users and the auditors of the RPA with a practical guide for banking action and thus ensure a safe and as far as possible supervisory-compliant use of RPA technology.

4.1 RPA as an Audit Topic—Widespread Use Requires Appropriate Auditing

The Federal Association of German Cooperative Banks (BVR) conducted a survey among its member institutes in 2020 on the potentials of process automation in cooperative banks (BVR, 2020). About 30% of the banks participated in the

M. R. Smeets et al., *Robotic Process Automation in Use*,
https://doi.org/10.1007/978-3-658-45902-4_4

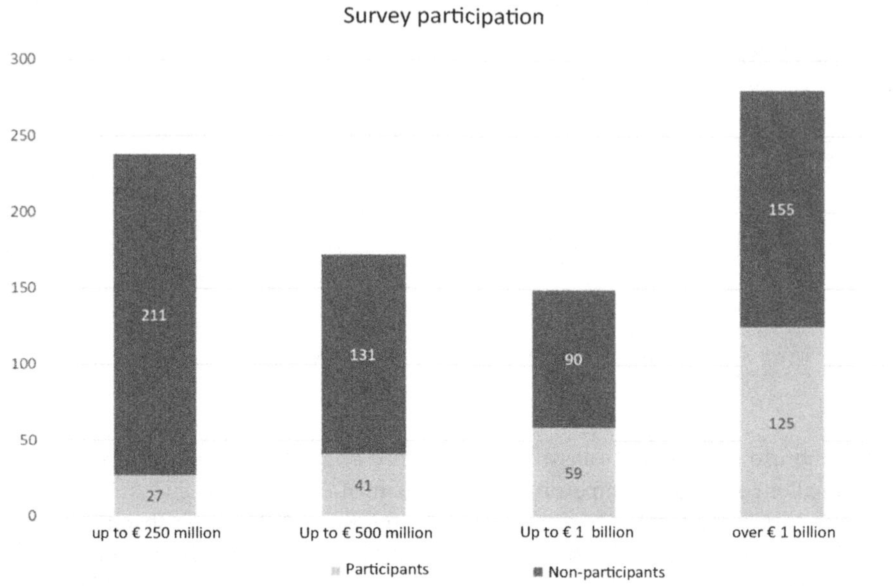

Fig. 4.1 Survey participation. (BVR, 2020)

BVR survey on process automation, Fig. 4.1. provides an overview of the survey participation.

The survey results thus provide a meaningful overview of practical application, especially for banks with an average balance sheet total of over 500 million €.

At that time, 7% of banks were already using RPA in day-to-day business, see Fig. 4.2, with increasing dynamics of activities (the share including explorers is 35%). Against this background, there is a clear commitment from the IT service provider of the cooperative financial group Atruvia AG to operate and promote process automation in the core banking process (Atruvia, 2022).

In 89% of cases, users rely on the standard software UiPath or UiPath Cloud. The specifics of this software or its eponymous provider and the terminology chosen will therefore be the focus of the analyses in the further course of the chapter.

Fig. 4.3 shows processes that were classified by the participating banks according to the criteria "importance" and "automation potential" (BVR, 2020).

RPA can basically be used in all business areas of the bank. The most important RPA use cases include the following (Ostrowicz, 2017 and Chap. 2):

Customer Service

RPA can improve customer service by automating tasks in the contact center, such as verifying electronic signatures, uploading scanned documents, and verifying information for automatic approvals or rejections. RPA not only standardizes and

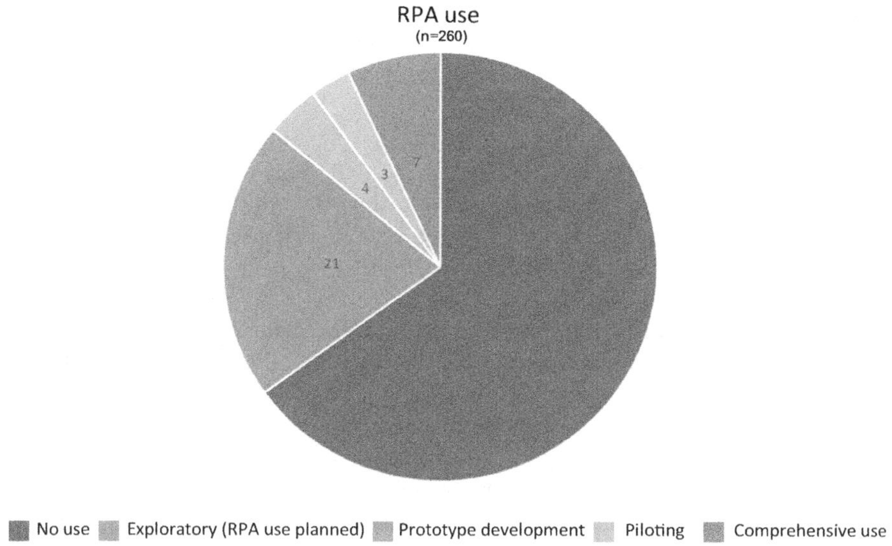

Fig. 4.2 RPA usage. (BVR, 2020)

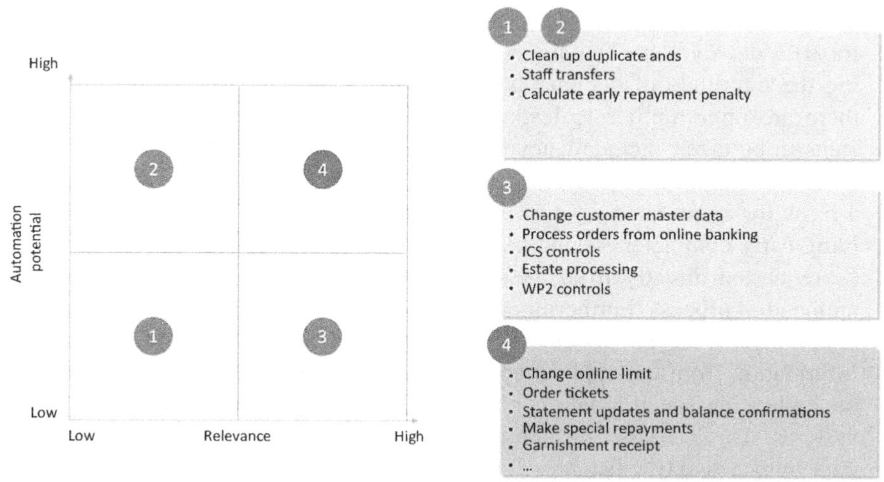

Fig. 4.3 Examples of processes suitable for RPA. (Based on BVR, 2020)

accelerates routine tasks, but also gives employees the necessary freedom to focus on solving customer problems.

Finance and Accounting

Financial and accounting processes suitable for RPA include general accounting, tax accounting and compliance, as well as financial planning and reporting.

Human Resources
Tasks in human resources that can be automated by RPA include payroll, time tracking and attendance management, as well as the structured recruitment of new employees.

IT Management and Services
Examples of RPA application in IT include the automation of software audits, source code control management, solving issues such as password resets and server restarts, and optimizing email notifications.

Furthermore, there are opportunities for RPA deployment *in almost every banking process*. Banking processes are often characterized by document-based input information. Automated document processing can deliver significant efficiency gains in terms of throughput times and quality. The faster processing of concerns and thus the delivery of a response to the submitted application almost in real time can ideally lead to an increase in customer satisfaction. Optimized document review can thus eliminate inefficiencies, seamlessly transfer data to subsequent processes and use it, and generate valuable net market time.

Example

Even in the complex real estate financing process, there is significant potential for RPA deployment (Douqué et al., 2022). Particularly in the phase of reviewing the multitude of documents and proofs to be submitted by the customer, there are opportunities to leverage optimization potential through RPA. The steps to be taken include categorizing the various submitted documents based on document-specific structures and patterns. This classification then serves as a basis for comparison against application-specific checklists using RPA, enabling early completeness and plausibility checks. Missing documents can also be requested directly from the customer in a timely manner to continue the application process. Furthermore, the documents can be archived automatically afterwards. Once all documents have been checked and are fully available, information from the application can finally be read out and customer-specific or application-specific data can be seamlessly transferred to internal banking systems. This can also include corresponding comparisons or enrichments of data with possibly existing information in the core banking system via RPA or the verification of signatures. The technical focus of the document review is then the comparison of the information provided in the application against defined technical rules. This can be optimized with its complex technical criteria using RPA to enable error-free and goal-oriented processing of potential applicants. ◄

▶ **Conclusion Important.** The wide range of applications for RPA and its steadily increasing use in practice require a thorough examination of the audit field of RPA (Hanenberg, 2001). According to BT 2.3 text numbers 1 and 2 of MaRisk, the activities of the internal audit must be based on a comprehensive and annually updated audit plan. The audit planning must be risk-oriented. The activities and processes of the institute, even if they are outsourced, must be audited at appropriate intervals, generally within three years. If there are special risks, an annual audit is required. For activities and processes that are not significant from a risk perspective, the three-year cycle can be deviated from. The risk classification of the activities and processes must be regularly reviewed. The risk assessment procedures of the internal audit must include an analysis of the risk potential of the activities and processes, taking into account foreseeable changes. The various sources of risk and the susceptibility of the processes to manipulation by employees must be appropriately taken into account (BaFin, 2021).

The topics discussed in this chapter can serve as explicit support and guidance for the planning to be carried out by the audit and bank against this background and the concrete preparatory and implementation measures derived from it and the project-accompanying or regular audit actions following in regular operation.

Example

Even the financial supervisory authority itself now uses automated assistance for extremely sensitive processes: The ECB banking supervision will in future use artificial intelligence (AI) to check candidates for leadership positions at banks. The European Central Bank (ECB) has launched the first version of a powerful natural language processing tool, named after a mythological god of the Vikings "Heimdall", ECB banking supervisor Elizabeth McCaul announced on 13.07.2022. The tool will support the ECB in its so-called "fit-and-proper" assessments. This is how the supervisory authority refers to the assessments of the professional qualification and personal reliability of candidates for leadership positions at supervised institutions. "Heimdall automatically processes and pre-analyses documents submitted by banks for potential leaders and key position holders," McCaul explained. Heimdall can read and process documents in natural language and structure the information according to predetermined rules for the supervisors. (FAZ, 2022) ◄

4.2 RPA and Audit—An Overview

4.2.1 Perspective RPA as an Audit Object

The use of RPA has fundamental immediate and indirect effects on the audit strategy and the audit activities of the audit department, especially with regard to the audit areas "control system" and "control activities". This will be illustrated below using the example of the automation of accounting-relevant processes and controls.

Example

This becomes transparent even with a standard audit field such as invoice receipt auditing: Human activities such as viewing invoices and matching them with order and delivery note can be traced and regularly leave a trail, e.g. entries in log files, accounting notes or release documentation. However, these traces are lost through the use of RPA. Invoices are picked up directly at the source at the mail receipt and then processed within the RPA. The result of the audit is then a "True" or "False". This decision is in turn linked to the next robot. The challenges for the auditor in this context are multi-layered: On the one hand, he must record the RPA environment and on the other hand, assess whether the robot reliably comes to a decision. ◄

Even for a highly specialized IT auditor, these challenges are anything but trivial. He can certainly bring his expertise to bear on technical questions regarding access rights, user administration and other IT-related aspects. But RPA is not just an IT issue, but also a fundamental process issue. The responsible auditors need special expertise regarding robot-supported process designs in order to be able to adequately review the use of RPA in their companies. Therefore, in addition to the IT auditor, the entire audit team is required and possibly also specialized external audit support is necessary to meet the numerous complex professional requirements regarding the structural and functional tests to be carried out in relation to the use of RPA.

▶ **Important** Due to their process-independent and holistic positioning, their special audit expertise regarding organizational structure and process, control systems and control procedures, the audit department can also create specific added value in connection with the introduction and use of RPA:

The audit department can support the company in integrating governance, risk, and control aspects in connection with the introduction of new technologies and can demonstrate opportunities for implementing automated controls in relevant business units and processes. The audit department can provide advisory support in navigating through each phase of the implementation of RPA.

The following aspects need to be addressed by the audit department in the individual phases of RPA implementation:

Before the Implementation of RPA

- Has the management defined a strategy, goals and priorities for the introduction of RPA throughout the company?

- What degree of automation currently exists? How mature are the processes in relation to occurring changes or the definition of new requirements?
- Have governance processes been updated to meet the additional risks of RPA?
- How is the redistribution and reprioritization of employees managed?
- Have all relevant functions within the company been involved early and continuously in the process?
- Does the company have the ability to deal with the increasing complexity of RPA?
- Have the aspects of security, compliance with legal and regulatory requirements, finances—including the internal controls of risk reporting—and audit as well as the impacts on people, processes and technology been considered?

During the Implementation of RPA

- Does the project plan for RPA implementation include appropriate milestones?
- Are design changes during implementation in line with the established framework for risks and controls in RPA governance?
- Are risks from third parties (e.g. contracts or licenses) considered in the selection of RPA products during implementation?
- Does the governance framework address the need for efficient and appropriate controls of the bots?
- How is coordination between the three lines of defense ensured?
- What mechanisms exist to monitor the development and effectiveness of the RPA platform based on company requirements?
- What does the problem-solving process look like, in which automation errors are promptly evaluated, corrected, tracked and communicated through a solution?

After the Implementation of RPA and Ongoing

- Does the framework for controlling risks and controls of the RPA still align with business strategies?
- How is RPA vendor management integrated into the company-wide system of service provider control—including the assessment of risks by third parties and software security?
- Have the impacts of RPA on the company been adequately considered in the multi-year audit plan?
- Are the employees well informed about the processes and controls for which they are responsible?
- What key performance indicators, including platform utilization and capacity management, trends in process automation and process (exception) recommendations, are being determined?
- How are the effectiveness and efficiency of RPA measured and reported compared to the created business case?
- Have additional opportunities been considered to use RPA for security, operation, control tests, audit functions, and cyber orchestration? (EY, 2018)

4.2.2 Perspective of RPA as Possible Audit Support

To improve its own effectiveness and efficiency, for example through cost reduction, increasing output quality or creating additional value for the company, the audit should consider the use of RPA. RPA can be used for audits in a variety of

- regular and frequently recurring
- structured
- less complex (i.e., there are clear business rules and a low or even no exception rate)
- digitally based

processes.

> **Important** Especially in activities such as transferring data from the banking system to an audit software, comparing data from different sources (e.g., from main or sub-ledger), extracting data (reading audit-relevant data from contract documents), processing and transforming data (converting data from spreadsheets, compiling overviews, etc.), RPA can also unfold its potential in audit activities.

In general, there are the following five fields of application for the use of RPA in audit practice (Internal Auditing, 2020):

Support
RPA can support the internal audit in carrying out various activities, such as data procurement. Instead of manually searching for transactions in the ERP system, saving them and later merging them with other data, RPA allows the audit to perform such tasks at the push of a button.

Validation
RPA can relieve the internal audit in the validation of data. For example, bots can examine the reported distances on business trips by automatically calculating the distance between the start and end location and comparing it with the information in the database.

Control Testing
RPA can support the internal audit in testing certain controls. A large number of processes can be represented as rule-based sequences of activities within a flowchart. With the help of RPA, deviations from processes or the corresponding control weaknesses can be automatically identified, for example by searching for duplicate invoices in accounts payable.

Data Generation
RPA can help the internal audit in generating data or enable the evaluation of new data. This includes, for example, data in PDF format or the evaluation of images, the manual evaluation of which would take a lot of time, but may be necessary to reach findings.

Reporting
RPA can relieve the internal audit in the creation of reports by automatically generating templates and filling them with information. Another application example is automatically generated reminders of follow-up measures.

Such use of RPA in the audit itself has various effects. For example, the staff is freed from monotonous click-copy-paste activities by the fully automated processing of audit tasks in the background, or RPA efficiently prepares more complex audit activities (structured reading of documents and collection of audit-relevant data). However, it is necessary to observe the various legal and regulatory requirements for RPA in the audit just as in the company itself before and during the use of RPA. Therefore, in Sect. 4.3, these legal and regulatory requirements are analyzed for relevance and the use of RPA is subsumed under these requirements.

4.3 Overview of Relevant Legal and Regulatory Requirements

4.3.1 General Requirements from the BGB

4.3.1.1 Contract Law

> **Important** Prior to the deployment of RPA, various contractual issues need to be clarified. Careful preparation in this area is particularly relevant when damage occurs during later use by RPA, for example, because data were incorrectly assigned or customer inquiries were answered incorrectly.

The first question that always arises is who must compensate for this damage. As a rule, the acting company will initially have to compensate for the damage, as it is in a direct legal relationship with the injured party. However, the company may be able to seek recourse if the damage was not caused by its own error. Whether and to what extent recourse is possible depends on how the underlying contract for the RPA software is structured. Here, three contractual situations are particularly relevant (Feldmann, 2022):

1. Standalone purchase of RPA software: The company acquires the software, implements and configures it itself.

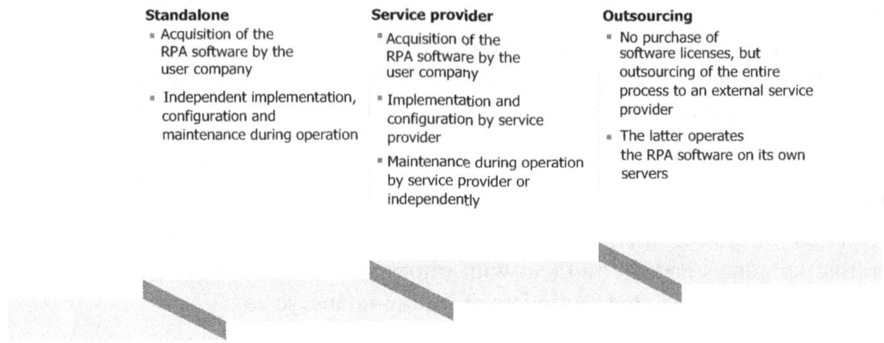

Fig. 4.4 Types of RPA usage. (Own representation based on Feldmann, 2022)

2. Purchase of RPA software and implementation by a service provider: Here, the company defines the requirements and the service provider implements them.
3. RPA outsourcing: The entire RPA process is outsourced to an external IT service provider, who operates the RPA software itself and integrates it into the company's processes.

The situations are graphically represented in Fig. 4.4.

If damage is now caused by the use of the RPA software, it must first be clarified who caused the damage. This depends on the design of the previously described contract situation. The liability risk for the deploying company is highest in the case of standalone purchase and decreases with increasing degree of outsourcing at the expense of the respective service provider. In practice, however, it is crucial that the injured party can not only claim the cause of the damage but also prove it. In principle, the claimant must prove that the damage was caused by a breach of duty by the defendant. For this to be possible, it is necessary to ensure from the outset in the RPA introduction process that the individual, automated process steps, tasks, and actions are documented in detail and can thus be verified.

The log files should provide information about whether errors have occurred or attempts have been made to manipulate the automated process from the outside. Likewise, the automated processes must be secured by a role and authorization concept so that only people who are authorized to do so can make changes. At the same time, it should be logged when which person made which changes. With a solid contractual sphere of influence in combination with appropriate organizational measures for documentation or evidence preservation in the course of service provision, legal risks from contract disputes can then be better managed in case of damage.

4.3.1.2 Tort Law/Damage Law

The complexity of RPA means that errors can occur in practical use.

Despite all the advantages that can arise from the use of RPA in the financial sector, the question arises as to who has the legal responsibility for an error or damage. Who is liable if a bot has answered a customer inquiry incorrectly, or if data has been assigned incorrectly?

Of course, the following statements do not replace a thorough, case-specific legal consultation, but they should provide basic information about existing and applicable principles of tort law:

The claims regulated in the BGB in the event of damage do not distinguish between human and machine errors. In practice, the company with which a victim has a business relationship is initially liable. In relation to the customer, this is initially the bank or financial service provider. However, a victim cannot just claim damage, they must also be able to prove and quantify it. They must prove that the damage was caused by misconduct or a breach of duty by the perpetrator. Here is an example from the investment sector: If the RPA-managed portfolio of a customer has lost a lot of value, a "damage" has occurred. However, a claim for damages would only be considered legally if the loss in value was not due to the volatility of the share prices, but the bot had invested the customer's deposits in an expressly excluded asset class. If it was contractually agreed that the customer's deposits would not be invested in cryptocurrencies, but this still happened, a court might consider a claim for damages (Behrens, 2021).

> **Important** Therefore, it is vitally important when implementing RPA processes to document the automated process steps, tasks, and actions in detail so that they can be reviewed.

Log files should contain clues as to whether an error has occurred or whether the process has been changed. Therefore, it is also very important that changes to the RPA processes can only be made by authorized users.

> **Important** If a damage is legally established, the RPA user has the option to make recourse claims, provided that this RPA was created by third-party service providers or it is not a completely own development.

Whether there are prospects of reclaiming one's own damage depends heavily on the design of the contractual relationship and would have to be checked on a case-by-case basis. Product liability law provides a number of bases for claims in the Product Liability Act, but at the same time also comparatively high hurdles of proof (Bonertz et al., 2020), which, however, should not be discussed in depth here.

4.3.2 Requirements of Commercial and Tax Law

Of course, when using RPA in accounting-relevant functional areas, the principles for proper management and storage of books, records and documents in electronic form as well as data access (GoBD) and all requirements of the Institute of Auditors (IDW) from the area of accounting, such as the IDW Statement on

Accounting: Principles of Proper Accounting when using Information Technology (IDW RS FAIT 1), the IDW Statement on Accounting: Principles of Proper Accounting when Outsourcing Accounting-Relevant Processes and Functions including Cloud Computing (IDW RS FAIT 5) and the C5 Criteria Catalogue (Cloud Computing Compliance Criteria Catalogue), must be observed (Dagianis, 2021).

For example, users often ask the following questions about how to interpret the GoBD when using RPA:

- What requirements are there for user authorization management?
- What guidelines are there regarding the appropriateness of control measures?
- Do access authorization concepts only need to exist, or are there specific requirements for these concepts? Are there concepts that would be insufficient?
- What guidelines are there regarding separation of functions?
- How are plausibility checks to be implemented? Are technical input controls sufficient?
- What guidelines are there regarding the appropriateness of protection measures against falsification?
- Who must report which errors detected during processing controls and when?
- What specific requirements are there for the registration process of new users?
- Is there a requirement for multi-factor authentication?
- What time intervals are appropriate for user account controls (e.g., in relation to seasonal suppliers)?
- When is the examination and processing of error messages and logs no longer timely?
- In what form must the service provider's mapping tables be available at the company?

The answers to these questions as well as certifications or proof of compliance with the aforementioned relevant IDW standards should be able to be provided by the service provider and are usually documented in corresponding white papers. Own audit actions by the user's revisions should therefore at least in this respect be able to be limited to the completeness and timeliness of the certificates provided.

4.3.3 Banking Requirements of the KWG

4.3.3.1 General Norm of the KWG

The supervisory general norm for the requirements of a proper business operation is § 25a KWG. This regulation stipulates that "an institution […] must have a proper business organization that ensures compliance with the legal provisions and business necessities to be observed by the institution. The managing directors are responsible for the proper business organization of the institution; they must take the necessary measures for the development of the corresponding internal institutional guidelines, unless the administrative or supervisory body decides. A proper business organization must in particular include an appropriate and effective risk management, [...]; risk management includes in particular [...]

- an appropriate personnel and technical-organizational equipment of the institution;
- the establishment of an appropriate emergency concept, especially for IT systems [...]" (Luz et al., 2011).

The interpretation of this general norm is done in the so-called minimum requirements. The various minimum requirements are so-called norm-interpreting administrative regulations, which represent a self-commitment of the German financial supervision towards the financial service providers (Wundenberg, 2012).

For banks, the minimum requirements for risk management (MaRisk) represent this central administrative regulation. The following presentations focus on the banking regulations, but due to the high redundancy with the minimum requirements for other sectors of the financial industry (such as insurance companies and investment companies), they are also relevant for these. The MaRisk concretize the § 25a KWG and are the implementation of the qualitative requirements from Basel II or Basel III for the risk controlling of banks and the corresponding banking supervisory review processes into German law (so-called "second pillar" of Basel II/III). The MaRisk are thus de facto a binding interpretation of the § 25a para. 1 KWG. They are intended to enable the supervisory authority to apply consistently to financial institutions or insurance companies and to create legal and planning certainty. Compliance with the MaRisk is audited by the auditor as part of the annual audit. They are also the subject of special audits according to § 44 para. 1 KWG. Such audits are carried out by auditors of the Bundesbank according to the new version of the supervisory guideline, which specifies the division of labor between BaFin and Deutsche Bundesbank on the basis of § 7 KWG. The use of RPA is also subject to these guidelines, the particularly relevant regulatory contents of which are examined in detail below.

4.3.3.2 Detailed Requirements MaRisk

4.3.3.2.1 AT 2.2 Risks/Risk Inventory

To assess the materiality of risks, the management board must regularly and on an event-driven basis obtain an overview of the institute's risks as part of a risk inventory (overall risk profile). The risks must be recorded at the level of the entire institute, regardless of which organizational unit caused the risks.

At least the following risks are generally to be classified as significant:

- Counterparty default risks (including country risks)
- Market price risks,
- Liquidity risks and
- Operational risks.

Risk concentrations associated with significant risks must be taken into account. Appropriate precautions must be taken for risks that are classified as not significant.

As part of the risk inventory, the institute must examine which risks can significantly affect the financial situation (including capital resources), the earnings situation, or the liquidity situation. The risk inventory must not be oriented solely towards the effects in accounting and formal legal designs (BaFin, 2021).

The basis for successful risk management is thus the systematic recording and analysis of all risks significant to the bank. As part of the risk inventory, banks identify, quantify, assess, and document their risks according to three dimensions:

- Financial situation (including capital resources)
- Earnings situation
- Liquidity situation.

The orientation of the risk inventory towards the three mentioned dimensions provides guidance regarding the consideration of a risk class assessed as significant in the risk management and controlling processes according to AT 4.3.2 MaRisk. The assessment of the influence on the financial situation has a strong connection to the economic perspective of the risk-bearing capacity (RTF) concept. The effect of a risk on the earnings situation dimension, and thus indirectly also on the future capital resources, is related to the impact on the normative perspective of the RTF. An assessment according to the liquidity situation clarifies whether and how the risk works in the ILAAP and is thus appropriately considered or integrated in the LTF.

The risk inventory thus forms the central building block in the context of overall bank and risk management (European Banking Authority, 2014).

To identify risks and cross-risk class effects at an early stage, banks have already derived suitable indicators today that ensure that the significant risks—also from outsourced activities and processes—can be identified early, fully recorded, and appropriately represented, so that no fundamental additions to the inventory system are necessary due to the use of RPA, but "only" appropriate consideration and integration is required.

> **Important** The use of RPA in the banking sector is primarily associated with operational risks. These are to be identified and described as part of the so-called risk inventory.

The term "risk inventory" emphasizes the need for a structured approach with appropriate documentation. This inventory must be carried out at least once a year or on an event-driven basis during the year. The risk manual, which as a framework document summarises the identified risks as well as key statements on the institute's risk management and controlling processes, is the commonly used method for documenting the inventory results. The inventory requires a holistic view according to the explicit expectations of the supervisory authority, which takes into account various influencing factors and goes beyond the balance sheet date. The materiality test is initially carried out using quantitative indicators that relate to the risk significance in relation to the risk coverage potential. In a second step, the quantitative materiality check is supplemented by qualitative considerations. Especially with regard to the second step, the use of RPA will have to be appreciated here as part of the risk inventory. For this second process step, BaFin has deliberately chosen the broad term "assessment" instead of "quantification" or "measurement" and thus follows the appropriate maxim "Better roughly and solidly justified right than precisely calculated wrong." (Hannemann et al., 2008):

Risks are explicitly defined as any form of deviation from the expected value, whether negative or positive (Gleißner & Romeike, 2005). The following risks, among others, can become significant when using RPA (Brettschneider, 2020):

- Incompatibility of third-party software with the core banking system and core banking processes, from the use of correct basic data to the storage of accounting, control or audit-relevant documents in the central document management system of the institution
- Systemic errors and individual errors in programming and the resulting consequences (damage cases, rework needs, production interruption)
- Risk of selecting unsuitable processes for RPA use (i.e., processes that are not sufficiently stable and automatable)
- Opportunity or renunciation of learning curve effects
- Cost risk (expenses for expanding the system landscape, managing access rights/process log maintenance, documentation effort, approval effort)
- Underestimation of know-how requirements (companies underestimate the IT expertise and infrastructure required for the maintenance and fine-tuning of RPA bots.)
- Due to digitalization, the number of interfaces between credit institutions and third parties is increasing significantly. This greatly increases the complexity of the IT infrastructure. This also enlarges and multiplies the potential attack surfaces that cyber criminals could exploit. This indirectly results in reputational risks, such as those arising from non-compliance with data protection regulations (University of Innsbruck, 2022)

The following manifestations of operational risk are likely to be regularly affected by the use of RPA (parcIT, 2022):

Information and Communication Technology Risk (ICT Risk)

ICT risk is defined as the risk of losses due to the unsuitability or failure of the hardware and software of technical infrastructures, which can impair the availability, integrity, accessibility, and security of these infrastructures or data. Here, the use of RPA, insofar as it substitutes the actions of employees, will initially have a slightly risk-increasing effect. However, it should be noted that employee-related process organization also has to organize numerous interfaces to hardware and software and ensure change management, so that overall risk neutrality should be achieved after successful implementation.

Behavioral Risk

Behavioral risk in institutions is understood as the risk resulting from the inappropriate provision of services, including damages caused by intentional or negligent behavior. Here, the use of RPA will have a significantly risk-reducing effect due to the automation effects.

Compliance Risk

Compliance risk is understood as the non-compliance with legal, supervisory regulations or internal regulations due to inadequate processes, negligent action or misinterpretation. Especially with the initial use of RPA, a number of external and internal guidelines have to be met. However, the statement on ICT risk applies analogously here, that with successful implementation, a risk-neutral effect will also be given in the context of risk inventory.

Other Operational Risks

Other operational risks are particularly those risks caused by external influences. These can be, for example, natural disasters, customer behavior, and legal risks without grandfathering (invalidity of contract conditions). These risks are generally independent of the use of RPA in their assessment.

In addition, there are so-called **cross-sectional risk classes**. These are to be considered across all risk classes.

In this context, the **reputational risk** could be particularly relevant. Reputational risk refers to the risk of potential damage to the institution's reputation in the public eye with effects on income and substance sizes. The risk often materializes in other risk classes, here in operational risks and specifically with regard to the consequences of non-compliance with data protection regulations. Here too, the use of RPA should not result in increased risks overall, as the systemic data protection risk due to the involvement of service providers is offset by the individual data protection risk due to employee failure.

> **Important** In summary, it can be stated that a cumulatively risk-relieving effect should be achievable through the targeted and systematically prepared use of RPA, particularly due to the positive effects on the probability of risk occurrence and risk controllability. At least an increase in risk is almost excluded with planned action during implementation and operation.

In a third and final step, the identified significant risks are to be analyzed for risk concentrations. The examination of risk concentrations takes place in the process of risk inventory for all risks assessed as significant. Risk concentrations must also be considered in view of the bank's earnings situation (earnings concentrations). In addition to risk positions vis-à-vis individual addresses, which represent a risk concentration solely due to their size, risk concentrations can arise both through the parallelism of risk positions within a risk class ("intra-risk concentrations") and through the parallelism of risk positions across different risk classes (through common risk drivers or through interactions of different risk drivers of different risk classes—"inter-risk concentrations") (Hofer, 2011).

> **Important** Here, the use of RPA should only be considered if modules of a provider, identical in terms of source code design, are widely used in various significant risk areas of the bank. Since the use of RPA in practice is still selective and highly specified for the respective field of application, risk concentrations are therefore likely not yet detectable here.

Looking at the near future, the following scenario should not be lost sight of: Thanks to new technical possibilities and due to competitive pressure, financial institutions increasingly outsource systems, activities or processes to industry-specific IT service providers in order to benefit from the cost-benefit advantages of scalable IT services. Many of these IT service providers, in turn, use the IT infrastructure of a common third-party IT service provider across several outsourcing levels. This could lead to a further concentration of IT infrastructure—for example, at cloud providers University of Innsbruck (2022). The risk concentration to be identified on this basis must then be addressed and analyzed in accordance with the requirements of MaRisk (see, for example, AT 4.2 para. 2, AT 4.3.2 para. 2, AT 4.3.3 para. 1).

4.3.3.2.2 AT 4.2 Strategies

In principle, any bank can use RPA immediately—without changing anything in its own structures.

> **Important** The probability of failure of such an RPA implementation without adequate preparation is, apart from the associated non-compliance with regulatory requirements and the resulting sanctions, almost certainly very high. The reason lies in the lack of compatibility between technology and the strategic "mindset" of the bank and its employees (Lomanto, 2019).

Introducing RPA does not just mean buying another software. RPA is a technology that must be strategically anchored in the deepest foundation of a bank, because RPA leads to processes not only being reconsidered, but usually completely rethought. This starts with small changes, such as a separate mailbox for incoming invoices, and leads to larger disruptions of previous habits. At the beginning of an RPA deployment, there should therefore be a strategic self-affirmation, in the course of which all processes and structures should first be reconsidered ("Re-Thinking") and checked for RPA capability based on the criteria listed below as examples (Allweyer, 2016):

- Stability of the process environment/process maturity/robustness of the process
- Degree of rule-basedness of the process
- Complexity of the process
- Transaction volume
- Requirements for the required data structures
- Frequency of repetition of the process
- System interface complexity
- Cost transparency of the process
- Degree of digitalization of the process
- Error risk of the process.

The processes selected for RPA deployment should then be adapted to the digital possibilities in the course of RPA development ("Re-Design").

> **Important** Against the background of this potentially far-reaching "Re-Thinking" and "Re-Design" of bank processes, intensive strategic preparatory work, the resulting derivation of strategic objectives, and the establishment of an adequate RPA governance are of crucial importance. The creation of a strategic "mindset" is therefore a minimum business and regulatory requirement for a successful RPA introduction in order to minimize the risk of failure.

This risk must be countered not only in the interest of the supervisors, but also in the bank's own interest through clear strategic guidelines:

Before an introduction, agreement should therefore be reached within the company on the following aspects and this should also be documented accordingly (Schmidt, 2020):

1. Vision

First, a company must develop a vision for RPA. What role should RPA play in the company? In which areas of the company should the technology be used and what kind of processes should be automated with it? These and other questions, which help to develop a vision of the role and function of RPA in the company, must be answered first. The answers already have implications for the second aspect.

2. Skills

Second, in order to successfully implement RPA, companies must have certain skills. Identifying, analyzing, and evaluating processes in terms of their automation potential requires appropriate knowledge about the possibilities, limitations, and functionality of the technology. In addition to developer and analyst skills, other skills such as project, test, and change management are also required. Companies must analyze what kind of knowledge and skills are already available and how they can provide the missing skills to the organization. Basically, there are two options: Either the company builds up the missing but necessary skills itself, or it buys them from external service providers. Both options have both advantages and disadvantages and must be weighed accordingly.

3. Organization

Third, companies must define who in the organization is responsible for RPA. The responsibilities and the structural anchoring of RPA in the organization are key success factors for the introduction and implementation of RPA projects. Companies can manage their automation projects centrally or decentrally. The project planning and development of RPA initiatives can also take place decentrally in the specialist departments, but also consolidated, for example in a corresponding process team with employees from various functional areas (Excellence-Center). Decisions about the organizational anchoring have a long-term effect on the scalability of RPA throughout the company and have a significant influence on the complexity and dependencies of future RPA projects.

Definition of success measurement indicators (Key Performance Indicators)

Finally, companies must measure the success of RPAs. It is absolutely necessary for companies to define metrics and processes for their collection and evaluation. Only in this way can companies determine whether RPA is actually successful. Ideally, the defined metrics check the progress with regard to the defined vision. Simply collecting the number of automated processes and determining more processes as a positive and fewer processes as a negative indicator does not do justice to the technology and its strategic aspects.

In accordance with the requirements of MaRisk, the management board must include these specifications in the sustainable business strategy of the institute, which presents the institute's objectives for each significant business activity and the measures to achieve these objectives. When defining and adjusting the business strategy, both external influencing factors (e.g. market development, competitive situation, regulatory environment) and internal influencing factors (e.g. risk-bearing capacity, liquidity, earnings situation, personnel and technical-organizational resources) must be taken into account (Hofer, 2011). Assumptions must be made with regard to the future development of the relevant influencing factors. The assumptions are subject to at least annual and event-related review; if necessary, the business strategy must be adjusted.

The management board must define a risk strategy that is consistent with the business strategy and the resulting risks. The risk strategy, possibly divided into sub-strategies for the significant risks, must include the objectives of risk control for the significant business activities and the measures to achieve these objectives. In particular, taking into account risk concentrations, the institute's risk appetite for all significant risks must be defined. Risk concentrations must also be taken into account with regard to the institute's earnings situation (earnings concentrations). This presupposes that the institute can distinguish its sources of success from each other and quantify them, e.g. with regard to the conditions and structural contribution in the interest book (BaFin, 2021).

Especially against the background of the aftermath of the pandemic, the current geopolitical, transformative and sustainable challenges as well as the constant regulatory new regulations, the (re-)evaluation of business models requested by the supervisory authority from a strategic point of view also requires intensive examination of the production processes. The supervisory authority expects calculations here also with regard to the economic viability of the business model. This definitely includes functioning and appropriate process cost accounting.

Various strategic priorities are conceivable in this context: Automation and RPA use contribute to the topic of process efficiency. With automation and the use of RPA, there is often a parallel reduction in the product range and/or product variants. Standardization generally also reduces process complexity. This in turn is a direct benefit that employees can experience.

Even with regard to the goal of customer focus, there are positive effects when making the strategic decision to increase automation: The degree of automation can be differentiated according to customer segments as needed. At the same time, appropriate automation for error reduction and thus indirectly also for improving customer satisfaction.

Below are some examples of strategic goal formulations in connection with the use of RPA that meet the requirements of the supervisory authority: "Our goal is to automate 50% of the productive activities of our bank by 2030."

Another exemplary strategic guideline could be to have the RPA cooperation partners guarantee and demonstrate consistently defined security standards of their RPA modules. These must not have any (negative) influence on the bank's data stocks and on the existing IT landscape.

> **Important** To fully prepare for the use of RPA, it is not only necessary from a regulatory perspective to anchor the topic in the business and/or risk strategy, but it is at least equally important to make the planned approach a cultural issue within the bank from the perspective of the company management. Because it is precisely the "Tone from the Top" that determines how the strategies are finally implemented in the bank (Glaser, 2019).

Finally, another strategic aspect in this context, at least in the view of BaFin, should be considered: The scenario described in the chapter "Risk Inventory" of a risk concentration with (cloud) providers in connection with RPA use could expand into a strategic risk for banks and their supervisors: Through the described outsourcing, cloud providers could access the data and business ideas of their customers, the supervised financial institutions. As a result, they could increasingly offer parts of the IT and financial services themselves. Thus, control could shift not only over the IT infrastructure, but also gradually over the financial services landscape to big tech companies. Banks and other financial institutions would thus be reduced to a risk-bearing shell, while the products and services would be produced in a non-regulated space. This highly undesirable risk from a supervisory perspective should be countered by early strategic decisions that should ensure as much autonomy as possible in terms of know-how and technical equipment at least at the level of the respective financial associations.

4.3.3.2.3 AT 4.3 ICS

In every institution, according to the nature, scope, complexity and risk content of the business activities

- Regulations for the organizational structure and processes must be made,
- Risk management and controlling processes must be established, and
- A risk controlling function and a compliance function must be implemented (BaFin, 2021).

When designing the organizational structure and processes, it must be ensured that incompatible activities are carried out by different employees and that conflicts of interest are avoided even when changing jobs. When employees from trading and market areas move to downstream areas and control areas, appropriate transition periods must be provided for activities that violate the prohibition of self-examination and self-review.

Processes and the associated tasks, competencies, responsibilities, controls and communication paths must be clearly defined and coordinated. Permissions and competencies must be granted according to the principle of economy (need-to-know principle) and adjusted promptly if necessary. This also includes the regular and event-related review of IT permissions, signing authorities and other granted competencies within reasonable deadlines. The deadlines are based on the importance of the processes and, in the case of IT permissions, the protection requirements of processed information. This also applies to the interfaces to significant outsourcing.

> **Important** In particular, the requirements for the organizational structure and processes require special attention and usually a higher implementation effort in the run-up to and environment of an RPA deployment.

These regulations and the design principles derived from them are therefore the focus of the following explanations:

A general requirement to be considered in the design of the organizational structure is an "appropriate and transparent corporate structure …, which aligns with the company's strategies and takes into account the transparency of the institute's business activities required for effective risk management …". This results from the overall responsibility of the managing directors for proper business organization (see § 25c para. 3 no. 4 KWG and AT 3 MaRisk). It is also required, among other things, that "as part of the organizational structure and processes, areas of responsibility are clearly delineated, with essential processes and associated tasks, competencies, responsibilities, controls and communication paths clearly defined and ensuring that employees do not perform incompatible activities" (see § 25c para. 4a no. 3 lit. a KWG).

The following specific requirements must also be ensured in connection with the use of RPA:

The more specific requirements for the organizational structure and processes are introduced in the MaRisk with BTO Tz. 1, which contains a general opening clause ("depending on the size, main business areas and risk situation"). The corresponding specifications are made in the text figures and explanations of BTO 1.1 and BTO 2.1. The general four-eyes principle described in AT 4.3.1 Tz. 1 for avoiding conflicts of interest is specified at various points in the MaRisk, for example in BTO 1.2.3 Tz. 1 for the credit process. Accordingly, process-dependent controls must be established for credit processing. Both functions (processing and control) can be located in one place, as long as the place consists of at least (these) two people.

The MaRisk specifically name the following positions:

- Accounting (BTO Tz. 7),
- Legal Department (BTO Tz. 8),
- Human Resources Department (BTO 1.1 Tz. 1 Explanation).

The "institutionalized four-eyes principle" required by the MaRisk in some text figures, for example, can be found in BTO Tz. 8, according to which significant legal risks must generally be checked by a body independent of the market and trading (e.g., the legal department). If the processing of credit and trading transactions, as in the case of RPA use, relies on IT systems in some areas, the separation of functions required by the MaRisk must also be reflected in the design of these systems. The applications at the IT level must not lead to a violation of the basic idea of function separation. This is to be ensured according to BTO Tz. 9 by implementing appropriate system-side procedures and protective measures. Such procedures and protective measures could, for example, include:

- Access security (e.g., password-protected use),
- Role and rights concepts, or
- Logging and evaluation of security and operationally relevant events (security monitoring).

Such security measures should be integrated into a comprehensive information security concept. In connection with the separation of functions and the clear definition of competencies and responsibilities, the MaRisk stipulate in AT 4.3.1 Tz. 2 that processes must be established in the institute for the regular and event-related review of

- IT authorizations,
- Signing authorizations, and
- Other granted competencies (e.g., trading or budget competencies)

The cycle for the regular review of the granted authorizations and competencies should be chosen depending on the risk associated with the authorizations. The standard here is the importance of the processes and, in the case of IT authorizations, the protection requirement of the information processed in the relevant systems. The focus is on possible damages that can result from the misuse of an incorrectly assigned authorization or competence. This can, for example, refer to the risk inventory according to AT 2.2 Tz. 1 or the determination of protection requirements according to AT 7.2 Tz. 4 MaRisk and the BAIT to assess the risk associated with a single authorization.

According to the explanation to AT 4.3.1 Tz. 2, a minimum cycle of a maximum of one year must be chosen for significant IT authorizations and signing authorizations in connection with payment transaction accounts. This also includes other authorizations that can be used for bookings or the release of bookings. For particularly critical IT authorizations, a routine review must be carried out at least every six months. Examples of particularly critical IT authorizations are, in addition to the administrator rights mentioned in the BaFin explanation, access to a trading system and so-called technical users. For other, demonstrably non-essential systems and other granted competencies, a review rhythm of up to three years is sufficient. However, this only applies to those authorizations and competencies from whose deliberate or accidental misuse no significant risks can arise. An event-related review of the authorizations and competencies is necessary, for example, when an employee takes on other tasks (see also AT 4.3.1 Tz. 1), or when there are changes within the IT systems (see also the later specific BAIT requirements).

Also in the development and change of programmatic guidelines for

IT systems (like RPA) separation of function aspects must be considered:

Every planning should carry out early tests that ensure both the technical functionality of the IT application or the IT system and the effectiveness of the IT security measures and their compatibility with the operation.

> **Important** The use of RPA in banks must be based on a regulated program release procedure (PRP). The scope and design of the PRP can be graded depending on the determination of protection requirements.

The tests (hardware and software) should be carried out in environments that are fundamentally separate from the production environment. The transition to production should be made through a final acceptance (release) that also takes into account the security aspects. The tests and the release must be documented accordingly.

> **Important** A standard process must be introduced for the tests and the release.

This also refers to the development and the implementation of the RPA into the production processes. Such a standard process is usually already established in the institute for economic reasons. The requirements are further specified in the BAIT (BaFin, 2017). Accordingly, the institute must regulate the guidelines for identifying all applications developed or operated by end users of the specialist department, for documentation, for programming guidelines and for the methodology of testing, for determining protection requirements and for the recertification process of authorizations, for example in an IDV guideline.

4.3.3.2.4 AT 5 Organizational Policies

The minimum requirements for the written order are set out in AT 5 of the MaRisk (BaFin, 2021):

The institution must ensure that business activities are conducted on the basis of written and publicized organizational policies (e.g., manuals, work instructions, or process descriptions). The level of detail of the organizational policies depends on the nature, scope, complexity, and risk content of the business activities, but must be so specific that it allows for an immediate audit.

The credit institution is free to determine the form of presentation. However, the level of detail of the organizational policies must be related to the nature, scope, complexity, and risk content of the institution's respective business activities. To meet the requirements for documented organizational policies, an institution can also use documents created by third parties (for example, descriptions of outsourced processes provided by a central service provider, or, as in the case of RPA use, certain process descriptions related to the software used). Externally created documents should be risk-oriented checked by the institution, adapted to individual circumstances if necessary, and possibly incorporated into its own organizational policies. A mere reference to the documentation within an external database or application, or to the documentation at the outsourcing company, is not sufficient, at least for essential regulations or determinations from a risk perspective, which should also include the regulations for RPA use.

> **Important** The institution should ensure that the presentation of the organizational policies is both appropriate and comprehensible to its employees. This means that facts, processes, and procedures must be specified. This is particularly challenging in connection with the technical specifics typical of RPA use.

The MaRisk set certain requirements for organizational policies. The policies must be written down, promptly adjusted in case of changes in activities and processes, made known to the affected employees in an appropriate manner, and made available to the employees in the current version.

The organizational policies must include the following:

- Regulations for the organizational structure and process, as well as for task assignment, competence order, and responsibilities,
- Regulations regarding the design of risk management and control processes,
- Regulations for the procedures, methods, and processes of aggregating risk data (for significant institutions),
- Regulations for internal audit,
- Regulations that ensure compliance with legal regulations and requirements (e.g., data protection, compliance),
- Regulations for procedures in case of outsourcing,

- Depending on the size of the institution and the nature, scope, complexity, and risk content of the business activities, a code of conduct for employees (BaFin, 2021).

In connection with RPA, after the strategic dimension, the regulation of responsibilities, and the structural anchoring of RPA in the organizational structure, special attention should be paid to the detailed description and regulation of the following interfaces:

- Design of the handover points from employee-based process parts to the RPA process part and back
- Design of the handover points of data and data sovereignty to external service providers and back (if relevant).

4.3.3.2.5 AT 6 Documentation

Business, control, and monitoring documents must be systematically and comprehensibly drafted for knowledgeable third parties and generally kept for five years. The timeliness and completeness of the record keeping must be ensured (BaFin, 2021).

The essential actions and determinations must be documented in a comprehensible manner. This also includes determinations regarding the use of significant opening clauses, which may need to be justified.

The opening clauses of the MaRisk give institutions a wide range of design options that strengthen their own responsibility. In addition, the requirements are very openly formulated, giving institutions room for institution-specific solutions. However, the institution-specific solutions often found in the use of RPA must ensure that the final and special auditors can get an idea of the institution's risk situation and the risk management and control procedures taken.

The business, control, and monitoring documents must be systematically and comprehensibly drafted and kept for knowledgeable third parties.

The basic retention period was raised from two years to five years in the course of the fifth MaRisk amendment on October 27, 2017. The five-year period corresponds to the regulation from § 25a para. 1 sentence 6 no. 2 KWG.143 From the addition "generally", it can be seen that specific regulations within the framework of other laws or regulations take precedence. Furthermore, the essential actions and determinations (for knowledgeable persons) must be documented in a comprehensible manner. This also includes determinations regarding the use of significant opening clauses, which usually need to be justified.

The documentation of essential decisions, processes, and procedures primarily serves to,

- coordinate the institution-specific regulations,
- provide the auditing bodies with insight into the institution-specific circumstances and decisions,

- establish a basis for assessing the proper business organization according to § 25a KWG and last but not least,
- facilitate the structured and efficient induction of new employees.

When creating and maintaining the documentation, the following principles should be considered:

1. All essential actions and determinations for fulfilling the MaRisk must be documented (see AT 6 Tz. 2).
2. The documentation must be comprehensible for knowledgeable third parties. This means that (technical and methodological) knowledge can be assumed.
3. It is not required that the documentation materials have the character of training or reference works. The form (electronic or physical), presentation (tabular, continuous text or graphic), and the depth of detail of the documentation are not prescribed by the MaRisk. Institution-specific solutions should be determined here, which also take into account the necessary resource expenditure.
4. Double documentation should be avoided. The recording of decisions, instructions, or dates only needs to be done in one place. If necessary, a reference to other documentation may be required for reasons of (overall) systematics. A meta-documentation (documentation about documentation) is not required by the MaRisk.
5. All business, control, and monitoring documents are (subject to legal regulations) generally to be kept for five years.

For practical implementation, it is recommended to adopt the legal retention periods into the work instructions relevant to the institution.

The following media can therefore serve to fulfill the documentation requirements:

- Strategies,
- Organizational guidelines and possibly framework conditions,
- Work or service instructions,
- Process descriptions,
- Manuals (e.g., risk manuals),
- Minutes of board decisions,
- Risk reports etc.

Regarding the content, the MaRisk are very openly formulated. As a rule, documentation will focus on the essential questions "What?", "When?" and "Who?".

What?

The "What?" specifies the content of the determination or decision. This includes, for example:

- Definition of definitions,
- Definition of parameters and assumptions

- Competence and escalation regulations
- Provisions for the separation of functions
- Definition and justification of the main risks.

When?
The MaRisk require institutions to individually determine dates, deadlines, and periods.

Who?
Finally, responsibilities must be defined within the framework of the MaRisk. This primarily concerns voting and decision-making powers in the credit business. But it must also be clearly regulated in other areas which body has to fulfill which tasks.

In connection with the highly specific considerations and determinations related to the use of RPA, it should be in the best interest of the applying institution to ensure complete documentation for each artifact over the entire life cycle from

- Idea
- Requirement catalog
- Concept
- Development
- Test
- Productive operation up to
- Decommissioning (Petersen & Schröder, 2020).

There should also be a prescribed standard procedure for this documentation itself. Special attention should be paid to sufficient separation of functions regarding the creation and approval of these documents.

The following audit questions about the organization must at least be answerable through the documentation:

- On what strategic basis does the development idea rest?
- Who is making the request?
- Is it purchased or developed in-house?
- Are there development guidelines?
- Is there a sufficient number of development users?
- How is testing done?
- Are there release procedures and how are they documented?
- Is there an implementation roadmap for working with mass data?
- Is a standard procedure defined?
- Is there a defined end of the artifact?

As part of risk management, the planned target requirements should be matched with the implemented security measures and remaining risks should be identified and managed appropriately. The decision on risk-bearing capacity is the responsibility of the technically responsible persons (Kokert & Held, 2013).

At the overall bank level, in particular to create the necessary transparency, the following supervisory expectation should be met by creating an appropriate documentation basis:

With a focus on information and communication technology, the Digital Operational Resilience Act (DORA) requires an information register with all contractual relationships associated with the use of ICT services. The term "outsourcing" is expanded by "ICT contractual relationships" and completes the sector map. DORA envisages the introduction of a comprehensive legal framework at EU level, which contains regulations for digital operational stability for all supervised financial institutions (European Commission, 2020). The proposal is currently being negotiated in the European legislative process and is expected to come into force during the course of 2023. It would be purposeful, also in the sense of creating one's own transparency, to create such a documentation basis already today out of one's own initiative.

The supervisory authority observes the fragmentation of value chains associated with the use of RPA rather skeptically and has formulated the following recommendations based on the findings of a research team from the University of Innsbruck to adequately address these challenges from their point of view (University of Innsbruck, 2022):

I. Create a comprehensive sector map

The research team from the University of Innsbruck recommends identifying the relationships and dependencies between financial institutions, IT service providers and other actors, as well as the concentration risks that arise from this. For this purpose, the supervisory authority should record all outsourcing (not only the essential ones) and fully include the subcontractors in the consideration—including their dependence on each other. Ideally, the supervisory authority also records the dynamic dependencies beyond the mere contractual relationship. This analysis could be presented in a sector map.

Such a sector map should also represent procedural dependencies (activity-based approach). To reduce complexity, the supervisory authority should uniformly specify the process definitions.

II. Work more closely with data protection authorities and competition supervision

The goals of IT supervision partly overlap with those of data protection authorities—for example, regarding security objectives such as confidentiality and integrity. The University of Innsbruck therefore recommends closer cooperation between the responsible authorities. The same should apply to the cooperation of BaFin with competition supervision. In this way, the authorities could better observe the increasing concentration of IT infrastructure and market power on individual companies and, if necessary, take measures when they identify systemic risks to financial stability.

III. Analyze actual use and vulnerabilities of PSD2 interfaces
From the perspective of the research team, there is already a tool that can be used to investigate how the supervised institutions interact with their service providers and what dependencies arise from this: the analysis of the use of interfaces, which have been created by the Second Payment Service Directive (Payment Service Directive—PSD2).

In addition, the research team recommends that the supervisory authority check the security of the certificates used by the payment service providers for the PSD2 interfaces—for example, by the BaFin accessing these interfaces with fake as well as with real, but revoked eIDAS certificates. The abbreviation eIDAS stands for electronic Identification, Authentication and Trust Services. Another idea: The supervisory authority could promote bug bounties—competitions in which participants go in search of vulnerabilities. Currently, third parties may be committing a criminal offense if they independently investigate interfaces.

> **Important** There is no disagreement from the industry's point of view with the described intentions and plans of the supervisory authority, as all planned measures contribute to transparency and risk reduction of RPA use and thus create the basis for an even broader use.

4.3.3.2.6 AT 7 Personnel/Technical-organizational equipment/ Emergency management

The quantitative and qualitative staffing of the institute must be oriented particularly towards internal operational requirements, business activities, and the risk situation. This also applies when resorting to temporary workers.

The employees and their representatives must have the necessary knowledge and experience depending on their tasks, competences, and responsibilities. Suitable measures must be taken to ensure that the qualification level of the employees is appropriate (BaFin, 2021).

Here too, the institute's strategy is initially the starting point and benchmark for the intensity and depth of the necessary employee competence. For institutes that use self-created RPAs due to their strategic orientation, for example, will have to meet higher standards in qualifying their personnel than institutes that use no or fewer "purchased" RPAs. This again demonstrates the principle of proportionality, which states that internal banking processes, and thus also the processes of personnel development or qualification, must be proportional to the size, business volume, and—as in the mentioned example—proportional to the risk structure (Deutsche Bundesbank, 2017).

In principle, the complexity of the businesses operated by the credit institutions requires an appropriate level of qualification of the employees and personnel planning that guarantees smooth operational processes at all points. This requirement, in particular, represents a not to be underestimated obstacle for smaller institutes for a more forceful use of RPA. Agile process teams, composed of employees from various functional areas (including process-accompanying auditing by the revision), can be a helpful tool for broadening the personnel base.

Explicitly required are qualification measures and knowledge and experience corresponding to the tasks, competences, and responsibilities in AT 7.1 Tz. 2 MaRisk. The requirements for the qualification of employees are subject to a constant process of change due to the rapid changes in the financial sector, this applies especially to the dynamic development in the field of RPA and AI. Therefore, the credit institution should create suitable measures to take this development into account. Suitable measures can be described as training and further education opportunities that correspond to the current or expected level of requirement for a particular job.

> **Important** Against this background, it is also sensible to develop a personnel development concept, which is oriented towards the operated or planned business and the associated functions and tasks. The creation of job descriptions for all areas of the credit institution is not necessary according to the MaRisk, but is highly recommended due to the high degree of specialization of the control staff relevant for the use of RPA.

In principle, personnel development does not take place once, but as a permanently ongoing process. AT 7.1 Tz. 3 also contains the requirement to operate a forward-looking deputy and successor development, so as not to be unprepared in the event of unexpected departure of important employees for the use of RPA (Spateneder, 2005).

> **Important** Although it is noted in the literature that a deputy regulation is not proportionate for all tasks, especially in small institutes (Helfer et al., 2008), the requirement should be taken into account with regard to the employees relevant for the use of RPA from a risk perspective. To meet this requirement, job descriptions can be drawn up or regularly updated, even if this is not required by the MaRisk.

Such job descriptions usually contain

- detailed descriptions of the tasks to be performed at the respective workplace,
- information about competences,
- information about responsibilities and
- superior and subordinate relationships (Berthel & Becker, 2003).

From these job descriptions, the requirements for the job holder can be derived, so that potential replacements can be sought for vacant positions with the existing profiles. In addition to this purpose, job descriptions and requirement profiles are also helpful in terms of personnel development, as a comparison of the defined requirement criteria with the knowledge and skills of the job holder can identify any need for further training.

In addition to personnel equipment, an equally high importance is placed on appropriate technical-organizational equipment and corresponding emergency precautions. Here, the supervisory requirements for emergency precautions will be discussed in more detail, while details on the technical-organizational equipment will be made in connection with the BAIT requirements:

The basis for a risk-oriented approach to emergencies is a proper identification of time-critical processes (Bretz, 2015). Precautions must be taken for emergencies in all time-critical activities and processes. The measures defined in the emergency concept must be suitable for reducing the extent of possible damage. Emergencies here do not only refer to the failure of IT systems, but also to the failure of essential business processes or locations.

Emergencies are delimited downwards to disturbances and upwards to disasters. The emergency can occur when a state is reached where the availability, integrity, or confidentiality of time-critical activities and processes cannot be restored within the required time, and this can result in high damage to the bank. The probability of occurrence is of secondary importance.

> **Important** For emergency preparedness according to AT 7.3 Tz. 1 of MaRisk, all measures to be taken after the occurrence of an emergency-triggering event, and all necessary information for this, must be documented in an emergency plan or emergency manual (type of emergency, consequences, measures to be taken and possible problems).

In the case of outsourcing time-critical business processes or other corresponding external procurement, the necessary emergency concepts of the institute and the cooperation company must be coordinated (BaFin, 2021). The institute therefore has the task, in the sense of appropriate risk management, to monitor the contract-compliant implementation of the requirements for restart at the service provider. A prerequisite for appropriate monitoring is that the service provider regularly confirms the effectiveness of its own emergency concept through tests.

The business continuity plans required in sentence 2 of AT 7.3 Tz. 2 should be created based on the business processes. Such business continuity planning helps to avoid business interruptions in advance through quick and efficient measures, or at least to ensure the continuation of company-relevant processes in an emergency. In business continuity planning, all activities are described to ensure a smooth flow of business processes even in emergency situations. All factors that can influence the process, such as infrastructure, IT systems, applications, personnel, documents, and service providers, must be taken into account.

Within the framework of the restart plans, based on criticality, emergency scenarios, and emergency resources, the restart time targets of the business processes are defined and the necessary measures are developed and documented. The service levels agreed with external service providers must also be taken into account.

The communication paths should be documented in the form of an alert plan, with the help of which the responsible persons or organizational units are informed in the event of an emergency. The alert can be made, for example, via telephone, fax, paging services, or courier. It must be described who notifies whom, who is to be notified as a substitute, or how to proceed if they cannot be reached. For this purpose, address and telephone lists may need to be kept.

> **Important** The aforementioned requirements for emergency preparedness are significant and contradict the common objective of RPA deployment, which is to achieve certain redundancies and replace them with a 365/24/7 available artifact. This can be pragmatically countered by initially limiting the use of RPA to the still very diverse non-time-critical fields of activity. This currently also seems to be compatible with practice in terms of the limitations of RPA use to fields of activity below a certain complexity threshold.

4.3.3.2.7 AT 8.2 Modification of operational processes or structures with ICS relevance

According to AT 8.2 MaRisk, the institute must analyze the effects of significant changes in the organizational structure and processes as well as in the IT systems on the control procedures and the control intensity and thus on the internal control system before they occur. The internal control system serves not only to meet legal requirements, but also continues to be a central management tool for supporting corporate objectives and reducing risk (Krekel & Faulmann, 2013). The organizational units that will later be integrated into the workflows must be involved in these analyses. The risk controlling function, the compliance function, and the internal audit must also be involved in their tasks (BaFin, 2021).

> **Important** Especially when using RPA modules for the first time, an analysis is strongly recommended to ensure and document the involvement of all relevant departments within the bank.

This is usually done in a standardized form. In addition to the objective, scope of use, planned deployment time, and specific description of the planned RPA use, the first materiality assessment of the planned change is then carried out. Indicators of materiality include the first-time use and the materiality of the business process affected by the introduction. Even if these two indicators are met, this is still not a precedent for conducting a detailed analysis. Further indications for determining the materiality of changes can be derived from the classification made in the area of IT security management.

If the need for a detailed analysis is denied, the insignificance of the effects on processes and/or the internal control system must be extensively justified qualitatively.

If a detailed analysis must be carried out, the effects of the RPA deployment on

- the functionality of the essential business process,
- the security and reliability of the data (very high protection requirement),
- compliance with the bank-specific legal norms essential for the bank (qualitative materiality criteria)
- and the risks arising from this for the result and capital (quantitative materiality criteria)

must be examined and documented.

In addition, the effects on the bank's internal control system must be examined. Here, the focus is also on reducing operational risks. Standardized work aids have also been developed by the respective associations of the credit industry for conducting the detailed analysis (Genossenschaftsverband-Verband der Regionen e.V., 2021).

A simplification rule can be applied: A detailed analysis can always be dispensed with if a comparable analysis has been carried out in the context of outsourcing or when acquiring the new software (Hannemann et al., 2019). This requirement will regularly be met when using RPA.

> **Important** Basically, when using RPA for the first time in sales, for example in the case of RPA-supported final sale of credit cards, a change process in the sense of AT 8.1 MaRisk can also be triggered.

The intention of this regulatory requirement is that every institute must understand the business activities it operates. For the commencement of business activities in new markets (including new distribution channels), a concept must be developed in advance. The basis of the concept must be the result of the analysis of the risk content of these new business activities and their effects on the overall risk profile. The concept must present the significant consequences for risk management resulting from this (BaFin, 2021). When conducting a comprehensive analysis in the sense of AT 8.2 MaRisk, the essential requirements of the change process in the sense of AT 8.1 MaRisk should be met in parallel. Here, the product and market catalog must finally be appropriately supplemented by the new distribution channel (RPA) (Hannemann et al., 2019).

4.3.3.2.8 AT 9 Outsourcing/Other External Procurement

The IT applications used when deploying RPA regularly fall under the provisions of MaRisk/BAIT if the processing results are relevant for accounting or control or if significant risks are associated with the use of the applications. There is discretion in this assessment (BaFin, 2019). Accounting relevance arises from the automated processing of data that is included in the accounting after processing. This also exists if balance sheet proofs are created using applications, provided no further evidence is available. Control relevance results from the processing of data, the results of which are used for significant business policy decisions or corporate control. Particularly relevant are evaluations that are used to meet the banking supervisory requirements of MaRisk-/BAIT (BaFin, 2021).

Further criteria relevant for the classification of the procured IT systems are:

Influence of the application on the customer relationship
The application or its processing results have a direct impact on the customer relationship. Thus, damages can occur for the bank, or a reputation risk is associated with the use.

Influence on Information Security
There is a high need for protection of the application with regard to the protection objectives of availability, integrity, confidentiality, and authenticity. For example, a failure of the application can have effects on the continuation of business operations, or there is a risk of high damage. Likewise, errors in the application can endanger the protection objectives.

Influence on Compliance with Legal/Regulatory Requirements
The application is necessary for the direct fulfillment of legal or regulatory requirements (outside the regulations of MaRisk/BAIT) (e.g., for evaluations for money laundering prevention or in reporting).

The decision-making process for categorizing the respective RPA application is schematically outlined as in Fig. 4.5 (BaFin, 2019):

> **Important** Regularly, the classification of the IT service as other external procurement should be appropriate for the currently used RPA software, see below.

For every other external procurement of IT services, a risk assessment must be carried out in advance. As far as IT services are already subject to a regular risk analysis as (essential or non-essential) outsourcing, no additional risk assessment as described below is necessary. The type and scope of the risk assessment for other external procurement of IT services are flexibly determined under proportionality aspects. Clustering of IT services for similar forms of other external procurement of IT services is useful, as existing risk assessments can be used for these.

A focus of the risk assessment for IT services are the information risks that can arise from other external procurement of an IT service. Therefore, the model risk assessment primarily focuses on the protection objectives of the business processes, data, and IT systems for which the IT service plays a role. Depth and scope of the risk assessment can differ depending on the need for protection. A new risk assessment is necessary when changes occur in the procurement of an IT service (in case of permanent procurement) and when the bank becomes aware of circumstances that suggest that the risk factors used in the assessment have changed (event-related risk assessment).

The risk assessment must be renewed every 3 years in case of permanent procurement or recourse to existing risk assessments for similar forms of other external procurement, even if there is no reason (regular risk assessment).

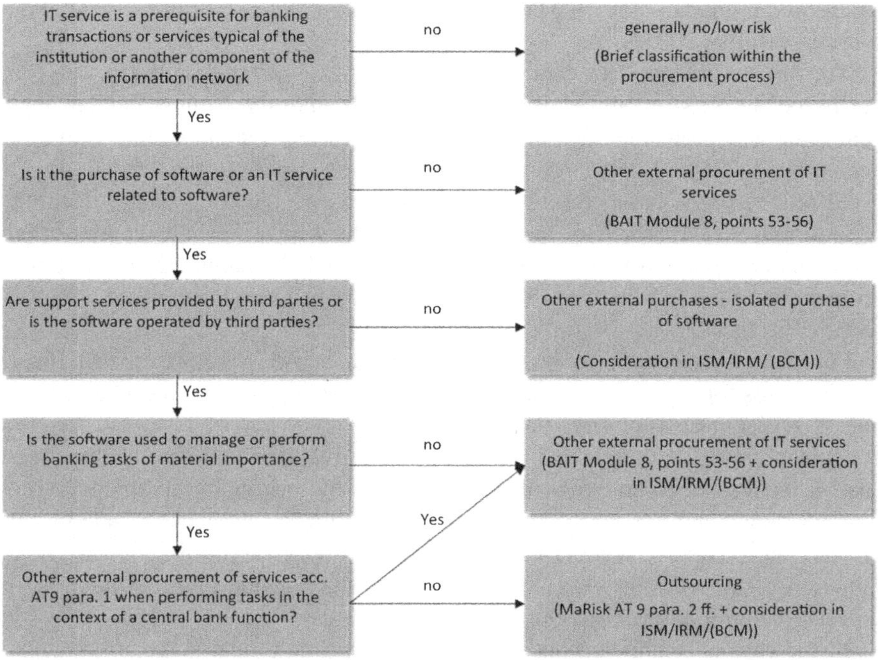

Fig. 4.5 Testing scheme for outsourcing/other external procurement. (BaFin, 2019)

In addition and as a precaution, however, against the background of the high importance attached to supervision, the requirements of the C5 criteria catalogue (Cloud Computing Compliance Criteria Catalogue) of the Federal Office for Information Technology (BSI) should also be examined for feasibility by the service provider in the case of other external procurement (Federal Office for Security in Information Technology, 2020). The criteria catalogue specifies minimum requirements for secure cloud computing and is primarily aimed at professional cloud providers, their auditors, and customers. The C5 criteria catalogue was first published by the Federal Office for Security in Information Technology in 2016, last updated in 2020, and is now considered a market standard. According to the BSI, more than a dozen attestations have already been created for national, European, and worldwide cloud providers as well as a wide range of cloud services. There are now also medium-sized and smaller providers who apply the catalogue. The C5 provides cloud customers with important guidance for selecting a provider. It forms the basis for carrying out a customer's own risk management. In 2019, the C5 was fundamentally revised to address current developments and further improve quality. The financial supervision also regularly refers to this criteria catalogue.

If, on the other hand, the bank comes to the conclusion in the decision-making process that the procured RPA IT service is relevant for bank control, the even stricter requirements of AT 9 MaRisk apply, even in the case of non-essential outsourcing (BaFin, 2021):

The institution must evaluate which risks are associated with an outsourcing based on a risk analysis. Based on this risk analysis, it must independently determine which outsourcings of activities and processes are essential from a risk perspective (essential outsourcings). This must be carried out regularly and on an event-related basis, based on uniform framework specifications at the institution or group level. The results of the risk analysis must be taken into account in outsourcing and risk control. The relevant organizational units must be involved in the preparation of the risk analysis. The internal audit must also be involved within the scope of its tasks.

For outsourcings that are not essential from a risk perspective, the general requirements for the proper organization of business operations according to § 25a para. 1 KWG must be observed.

Outsourcing of activities and processes in control areas and core banking areas can be carried out to an extent that ensures that the institution continues to have the knowledge and experience to effectively monitor the services provided by the outsourcing company. It must be ensured that, if necessary—in the event of termination of the outsourcing relationship or changes in the group structure—the orderly operation in these areas can continue.

In the case of significant outsourcing, the institution must make arrangements in the event of intended or expected termination of the outsourcing agreement to ensure the continuity and quality of the outsourced activities and processes even after termination. For cases of unintended or unexpected termination of these outsourcings, which may be associated with a significant disruption of business operations, the institution must examine and adopt any possible courses of action for their feasibility. This also includes, as far as sensible and possible, the definition of corresponding exit processes. The courses of action must be reviewed regularly and on an event-related basis.

In the case of significant outsourcing, the following must be agreed in particular in the outsourcing contract documented in text form:

- Specification and possibly delineation of the service to be provided by the outsourcing company,
- Date of the beginning and possibly the end of the outsourcing agreement,
- If deviating from German law, the applicable law for the outsourcing agreement,
- Locations (i.e., regions or countries) where the service is performed and/or relevant data is stored and processed, as well as the regulation that the institution will be notified if the outsourcing company changes location,
- Agreed service quality with clearly defined performance targets,
- If applicable, that the outsourcing company has to provide proof of insurance for certain risks.
- Requirements for the implementation and review of emergency concepts,

- Establishment of appropriate information and audit rights of the internal audit and external auditors,
- Ensuring unrestricted information and audit rights as well as control options of the authorities responsible according to § 25b paragraph 3 KWG regarding the outsourced activities and processes,
- If necessary, rights of instruction,
- Regulations that ensure compliance with data protection regulations and other security requirements,
- Termination rights and appropriate notice periods,
- Regulations on the possibility and modalities of further outsourcing, which ensure that the institution continues to comply with the banking supervisory requirements,
- Obligation of the outsourcing company to inform the institution about developments that could affect the proper execution of the outsourced activities and processes.

With regard to further outsourcing, the outsourcing contract should, as far as possible, include the outsourcing institution's right of approval or specific conditions under which further outsourcing of individual work and process steps is possible. At the very least, it must be contractually ensured that the agreements of the outsourcing company with subcontractors are in line with the contractual agreements of the original outsourcing contract. Furthermore, the contractual requirements for further outsourcing must also include an obligation of the outsourcing company to inform the outsourcing institution. It must be ensured that the outsourcing company remains accountable to the outsourcing institution in the event of further outsourcing to a subcontractor.

The institution must adequately manage the risks associated with outsourcing and properly monitor the execution of the outsourced activities and processes. This includes, in the case of significant outsourcing, the ongoing monitoring of the performance of the outsourcing company based on criteria to be maintained (e.g., Key Performance Indicators, Key Risk Indicators) and contractually agreed information from the outsourcing company; the quality of the services provided must be regularly assessed.

For the documentation, control, and monitoring of significant outsourcing, the institution must define clear responsibilities. To meet these high requirements at the institutional level with the necessary quality standard and the expertise required especially for IT services, many cooperative banks, for example, use a specialized central service provider, to whom the central outsourcing management is then transferred.

The requirements for the outsourcing of activities and processes must also be observed in the case of further outsourcing of outsourced activities and processes.

In principle, the institution must maintain a current outsourcing register with information about all outsourcing agreements. The minimum content requirements for the outsourcing register can be found for all outsourcings in Tz. 54 and for significant outsourcings in Tz. 55 of the EBA guidelines on outsourcing (European

Banking Authority, 2019). The outsourcing register includes all outsourcing agreements, including the outsourcing agreements with outsourcing companies within an institution group or a financial association. Furthermore, in the case of further outsourcing of significant outsourcings, the outsourcing institution must determine whether the part to be further outsourced is significant and this significant part must be recorded in the outsourcing register.

In terms of financial associations, certain simplifications arise (see AT 9 Tz. 15 MaRisk, BaFin, 2021).

The intentions possibly originally associated with the use of RPA

- Better control, as the institution itself regulates process automation
- Cost advantage (RPA conception and implementation usually cause lower administrative costs than an outsourcing process)
- Greater versatility and individualization (no rigid performance records)
- Scalability through AI
- Higher availability (365/7/24)

are at least partially abandoned when classified as outsourcing.

> **Important** In practice, the processes operated with RPA are also likely not to be relevant to bank control due to the (still) often low complexity level. Against this background, a further in-depth examination of the regulatory requirements for outsourcing is waived here for the time being.

4.3.3.3 BAIT Detailed Requirements

4.3.3.3.1 IT Strategy

The IT strategy must meet the requirements according to AT 4.2 of MaRisk. This includes in particular that the management defines a sustainable IT strategy in which the goals and measures to achieve these goals are presented.

The management must define an IT strategy consistent with the business strategy. Minimum contents are:

- Strategic development of the IT structure and IT process organization of the institute as well as IT services and other important dependencies on third parties
- Assignment of common standards, which the institute follows, to the areas of IT and information security
- Goals, responsibilities and integration of information security into the organization
- Strategic development of IT architecture
- Statements on IT emergency management taking into account information security concerns
- Statements on the IT systems (hardware and software components) operated or developed in the departments themselves (BaFin, 2017).

As already described above, at the beginning of the decision-making process about using RPA, there is usually a strategic self-assurance, in the course of which all processes and structures should first be reconsidered and checked for RPA capability. If then in the further course the strategic decision for RPA use is made, strategic goals and metrics are defined and finally operationalized, a fundamental architecture decision must be made:

The entry into RPA use often starts with a so-called Attended (stands for: supervised) Robot. This type of robot is a "supervised" robot. Robots of this type run locally on a physical computer such as a laptop or workstation. The robots are programmed here to work in collaboration with the user.

In contrast, the use of so-called Unattended Robots is only possible via a networked platform. The Unattended Robot is therefore an "unsupervised" or "autonomous" robot. This can work completely autonomously without the supervision of a human. This usually no longer happens on a physical machine, but on a virtual machine in a data center. This type of robot also needs a kind of "control center" which controls the bot. In UiPath, this instance is called "UiPath Orchestrator" (UiPath, 2022). See also Chap. 2 and Fig. 4.6, which summarizes the two different operating models.

Another central determination concerns the degree of centralization of RPA use. Here, three models are basically distinguished:

Central Model

Here, the RPA roles and tasks are in their own department, which has control over all RPA robots of the company. As a result, one usually benefits from short communication paths within the RPA team, has a clear overview of all automated processes and concentrated RPA knowledge. The risks are potentially less agility in the individual departments and the risk of neglecting individual areas of application due to central resource bottlenecks.

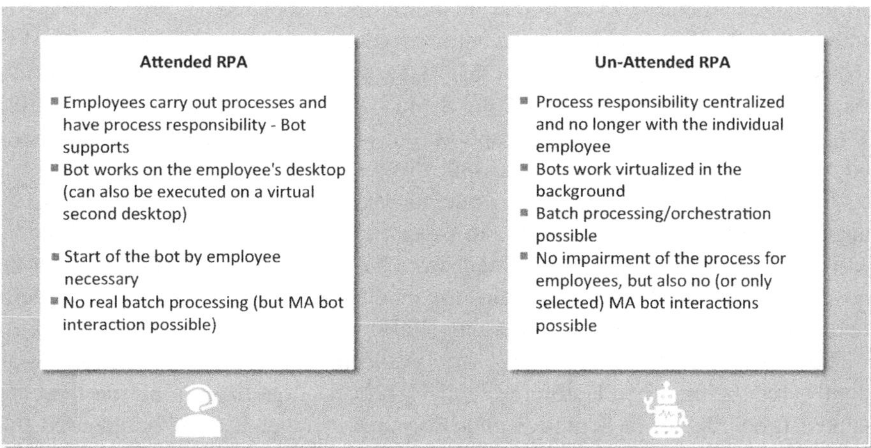

Fig. 4.6 Attended and Unattended RPA. (Own illustration)

Decentralized Model

The RPA units are each anchored in the individual departments and make their own decisions here. This can affect the tools used, processes to be automated, or the applicable guidelines themselves. Advantages are flexibility in development and direct communication channels between the RPA unit and the department heads. Disadvantages can be seen in the redundant structure of RPA roles and infrastructures, possibly higher costs, and potentially lacking company-wide scalability.

Hybrid Model

The hybrid combines some advantages of the central and decentralized models. It offers both a strong anchoring of RPA in the organization and high flexibility in the development and maintenance of RPA bots. In addition, there is a sufficient pool of RPA-experienced individuals available for the same platform and the direct communication channels between those who develop the RPA processes and those who use them in the department remain intact. However, the challenge here is that the individual departments are continuously trained in terms of the specifications of the central RPA governance. There is also increased coordination effort to ensure institution-wide consistency of automation approaches (Dagianis, 2021).

The three approaches are summarized in Fig. 4.7.

> **Important** Depending on these two aforementioned basic determinations, the extent of the supervisory requirements described below is also shaped. The architecture of the application landscape must therefore be supplemented in accordance with the requirements of the BAIT.

4.3.3.3.2 IT Governance

IT governance is the structure for controlling and monitoring the operation and further development of IT systems, including the associated IT processes, based on the IT strategy. Particularly relevant are the regulations for IT structure and IT process organization (see AT 4.3.1 MaRisk), for information risk and information security management (see AT 4.3.2 MaRisk, AT 7.2 Tzn. 2 and 4 MaRisk), for quantitatively and qualitatively appropriate staffing of IT (see AT 7.1 MaRisk), and for the scope and quality of technical-organizational equipment (see AT 7.2 Tz. 1 MaRisk). Regulations for IT structure and IT process organization must be adapted promptly when activities and processes change (see AT 5 Tzn. 1 and 2 MaRisk). In principle, this requirement from Tz. 2 of the BAIT will be met if the above-described MaRisk requirements are met. However, the BAIT further specify the catalog of requirements considerably (BaFin, 2017):

The management is responsible for ensuring that, based on the IT strategy, the regulations for IT structure and IT process organization are defined and adapted promptly when activities and processes change. It must be ensured that these regulations are effectively implemented. Especially with (initial) RPA

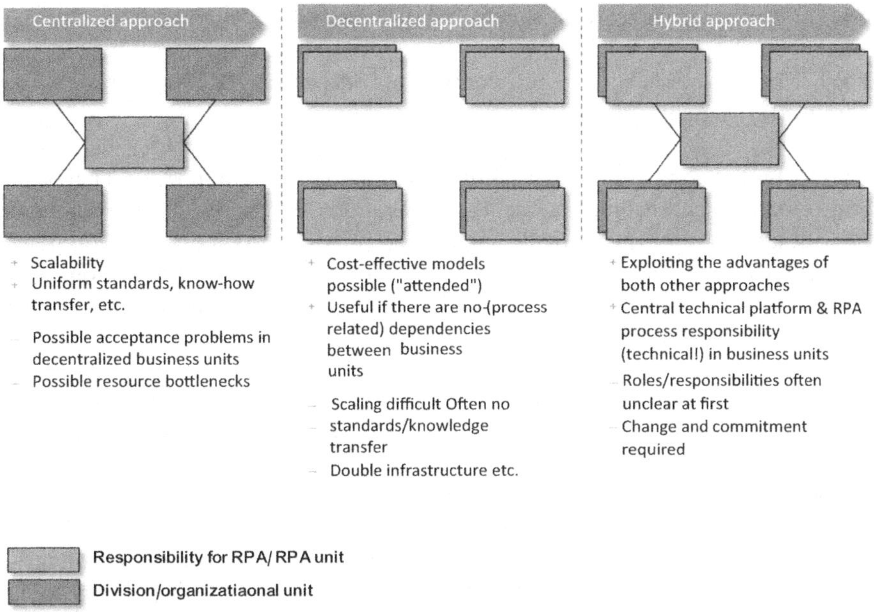

Fig. 4.7 Degree of centralization of RPA use. (Own illustration)

implementations, the supervisory authority expects a proactive creation of an appropriate structural and procedural framework. Not the artifact itself poses the risk, but its technical input and operation. With an increasing number of artifacts, some risks and challenges arise:

Lack of standardized guidelines and regulations for the procedure make targeted and consistent process automation difficult. Existing potentials are often not identified and used. This deficiency also carries the risk of inadvertently violating legal and regulatory requirements. The lack of definition of roles and responsibilities leads to potential conflicts or conflicts of interest. Effective knowledge preservation and regulated knowledge transfer are also difficult if existing expertise is not fully shared across the company and cross-departmental synergy effects are not recognized.

A specified RPA governance is highly relevant for achieving company-wide goals and securing business success. Proper RPA governance answers, among other things, the following questions:

- Who decides which processes are automated?
- Which technology, which provider, and which infrastructure are best suited?
- What roles are needed for the proper development of robots?
- Is a separation of functions ensured, if necessary?
- In what way and by whom are the robots controlled?

- What activities are carried out by IT, the specialist department, the process team, etc.?
- What organizational forms are conceivable (decentralized vs. hybrid vs. centralized) (Dagianis, 2021)?

> **Important** The institute must particularly equip information risk management, information security management, IT operations, and application development quantitatively and qualitatively appropriately with resources. The equipment, especially with financial and personnel resources, must be proportional, particularly with regard to the selected RPA architecture (attended robot or unattended robot) and the organizational form.

Conflicts of interest between activities, for example in connection with application development and the tasks of IT operations, must be addressed by structural or procedural measures or by an adequate role definition. Appropriate quantitative or qualitative criteria must be defined by the management for controlling the areas responsible for the operation and further development of the IT systems. Compliance with the criteria must be monitored. When defining the criteria, the quality of service delivery, availability, maintainability, adaptability to new requirements, security of the IT systems or the associated IT processes, and their costs can be taken into account.

> **Important** This permanent monitoring requirement naturally also applies to the use of RPA.

Control and monitoring measures are required both by the operational area and the function holders of the so-called "second line of defense" (here especially IT security officer, data protection officer, and MaRisk compliance function). In addition, the audit is also required to establish appropriate audit measures. In practice, for example, a permanent project-accompanying audit with regular reporting obligation could be an appropriate implementation suggestion. In addition, it is useful to integrate the use of RPA into the existing audit field, depending on the field of application, in order to avoid redundancies as far as possible.

4.3.3.3.3 Operational Information Security

Operational information security implements the requirements of information security management. IT systems, the associated IT processes, and other components of the information network must ensure the integrity, availability, authenticity, and confidentiality of the data. For these purposes, the design of the IT systems and the associated IT processes should generally be based on common standards (see AT 7.2 Tz. 2 MaRisk, BaFin, 2021). Appropriate monitoring and control processes for IT risks must be established, which include the definition of IT risk criteria, the identification of IT risks, the determination of the need for protection, derived protective measures for IT operations, and the definition of appropriate measures for risk treatment and reduction (see AT 7.2 Tz. 4 MaRisk).

In detail, this means

The institute must implement appropriate operational information security measures and processes that are in line with the state of the art, based on the information security policy and information security guidelines.

Threats to the information network should be identified as early as possible. Potentially security-relevant information should be evaluated in a timely, rule-based, and centralized manner. This information must be protected during transport and storage and be available for an appropriate period for later evaluation.

An appropriate portfolio of rules for identifying security-relevant events should be defined. Rules should be tested before being put into operation. The rules should be regularly and event-relatedly checked for effectiveness and further developed.

Security-relevant events should be analyzed promptly, and appropriate responses to resulting information security incidents should be made under the responsibility of information security management.

In coordination with the selected service provider, the specific measures for information security for RPA use should be initially agreed upon and continuously evaluated and refined based on the findings from practical use.

The security of IT systems should therefore be regularly, event-relatedly, and without conflicts of interest checked. Results should be analyzed for necessary improvements and risks should be managed appropriately. For the audit, these analysis results are indications for the continuous review of their internal risk assessment and possibly a reason for audit plan adjustments (Kurowski, 2004).

> **Important** To implement these requirements, the leading associations of the German banking industry have designed aids for the banks, which can also integrate aspects relevant to RPA use (Federal Association of Cooperative Banks, 2020).

4.3.3.3.4 Identity and Rights Management

Identity and rights management ensures that the permissions granted to users are designed and used in accordance with the organizational and professional guidelines of the institute. Identity and rights management must meet the requirements of AT 4.3.1 Tz. 2, AT 7.2 Tz. 2 and BTO Tz. 9 of MaRisk. Any access, entry and admission rights to components or parts of the information network should be subject to standardized processes and controls (BaFin, 2017).

Especially from the perspective of an auditor and the revision, it should be noted that the robot as an independent user regularly has to receive extensive rights in order to be able to perform its tasks. In addition, numerous rights must also be granted to bank employees and external parties in the context of implementing RPA. Their integration into the bank's identity and rights management is therefore essential.

Authorization concepts define the scope and terms of use of the permissions for the IT systems (access to IT systems and access to data) as well as the access rights to rooms consistently with the determined protection requirements and completely and traceably derivable for all provided permissions.

Authorization concepts must ensure the granting of permissions according to the principle of economy ("need-to-know" and "least-privilege" principles), maintain the separation of functions also across authorization concepts and avoid conflicts of interest. Authorization concepts are to be regularly and event-related reviewed and updated if necessary.

Accesses and entries must always be clearly assignable to an acting or responsible person (preferably automated).

The procedures for setting up, changing, deactivating or deleting permissions for users must ensure through approval and control processes that the specifications of the authorization concept are adhered to. The technically responsible body must be appropriately involved so that it can fulfill its technical responsibility.

When checking whether the granted permissions are still needed and whether they comply with the specifications of the authorization concept (recertification), those responsible for setting up, changing, deactivating or deleting permissions are to be involved Control instances (Becker, 2022).

The establishment, modification, deactivation and deletion of permissions and the recertification are to be documented in a comprehensible and evaluable manner.

The institute must establish processes for logging and monitoring in accordance with the protection requirements and target requirements, which make it verifiable that the permissions are only used as intended. Due to the associated far-reaching intervention possibilities, the institute must in particular establish appropriate processes for logging and monitoring for activities with privileged (particularly critical) user and access rights.

Technical-organizational measures should prevent the bypassing of the specifications of the authorization concepts (BaFin, 2017).

According to the BAIT, authorization concepts represent the central component of authorization management (Bretz, 2015).

> **Important** Since permissions in connection with the use of RPA are often interdependent in practice, the permission chain must always be considered closed and consequently the need-to-know principle must also be implemented in the context of these permission chains (Tsolkas et al., 2010).

In this context, it is also advisable to develop appropriate test scenarios for permission tests (Koenen, 2016).

In audit examinations, the following findings are often made on this subject (Conrads et al., 2015):

- Actual access rights deviate from the organizational assignment
- Insufficient assurance of the separation of functions and avoidance of conflicts of interest

- Insufficient target role concept
- Access rights to sensitive data are too extensive
- Insufficient rights granting process for privileged user IDs
- Rights are granted directly to individual users despite the role concept
- Service provider permissions are granted in a non-transparent manner
- Insufficient implementation of recertification.

4.3.3.3.5 IT Projects and Application Development

Both regulatory aspects are relevant in connection with RPA. In particular, the (initial) introduction of RPA is likely to meet the requirements of an IT project. Similarly, the creation of bots is regularly classified as application development. Therefore, the contents of both "requirement packages" must be implemented:

Significant changes in IT systems within the framework of IT projects, their impact on the IT organizational structure and IT processes, are to be evaluated within the framework of an impact analysis (see AT 8.2 Tz. 1 MaRisk). With regard to the initial use and significant changes of IT systems, the requirements of AT 7.2 (especially Tz. 3 and Tz. 5) MaRisk, AT 8.2 Tz. 1 MaRisk and AT 8.3 Tz. 1 MaRisk must be met (BaFin, 2017).

The organizational foundations for IT projects and the criteria for their application must be regulated.

IT projects must be appropriately managed considering their objectives and risks in terms of duration, resources, and quality. For this purpose, procedural models must be defined, and their compliance must be monitored.

The portfolio of IT projects must be appropriately monitored and managed. It must be taken into account that risks can also result from dependencies between different projects.

Significant IT projects and IT project risks are regularly and event-related reported to the management. Significant project risks must be taken into account in risk management.

Application development is comprehensively regulated in the BAIT (BaFin, 2017):

Appropriate processes must be defined for application development, which include specifications for requirement determination, development goal, (technical) implementation (including programming guidelines), quality assurance, and testing, acceptance, and release. Application development includes, among other things, the creation of software for business and support processes (including individual data processing—IDV). The design of the processes is risk-oriented.

With a view to the financial services sector, a well-established phase model is often used here (Claaßen, 2015). This model is divided into four phases:

- Analysis
- Design
- Development and Testing
- Implementation.

It is important to define exact responsibilities for each phase of this model.

The content requirements for the individual phases are then defined in detail in the BAIT (BaFin, 2017). The (extended) phase model described in Chap. 2 takes into account the aspects outlined below:

Requirements for the functionality of the application must be collected, evaluated, documented, and approved, as must non-functional requirements. Acceptance and test criteria must be defined for each requirement. The responsibility for the collection, evaluation, and approval of the technical requirements (functional and non-functional) lies with the technically responsible bodies. Requirement documents can differ depending on the procedural model and include, for example:

- Functional concept (specification)
- Technical functional concept (requirements specification)
- User-Story/Product Back-Log.

Non-functional requirements for IT systems include, for example:

- Requirements for information security
- Access regulations
- Ergonomics
- Maintainability
- Response times
- Resilience

As part of application development, appropriate precautions must be taken depending on the level of protection required, so that the confidentiality, integrity, availability, and authenticity of the data to be processed can be reliably ensured even after each application is put into production. Suitable precautions include, for example:

- Checking the input data
- System access control
- User authentication
- Transaction authorization
- Logging of system activity
- Audit trails (Audit Logs)
- Tracking of security-relevant events
- Handling of exceptions.

The integrity of the application (especially the source code) must be appropriately ensured. In addition, precautions must be taken, among other things, to detect whether an application has been accidentally changed or deliberately manipulated. A suitable precaution, taking into account the level of protection required, can be the review of the source code. The review of the source code is a methodical investigation to identify risks.

The application and its development are to be documented in a clear and comprehensible manner for knowledgeable third parties. The documentation of the application includes at least the following contents:

- User documentation
- Technical system documentation
- Operational documentation.

For example, versioning of the source code and requirement documents contributes to the traceability of application development.

A methodology for testing applications before their initial deployment and after significant changes is to be defined and implemented. The tests must include the functionality of the application, the implemented measures for information protection, and, if relevant, system performance under various stress scenarios. The technically responsible bodies are responsible for carrying out acceptance tests. Test environments for conducting acceptance tests must correspond to the production environment in aspects essential for the test. Test activities and test results are to be documented. The execution of tests requires relevant expertise of the testers and an appropriately designed independence from the application developers. The protection requirements of the data used for testing must be taken into account.

A test documentation includes at least the following points:

- Test case description
- Documentation of the underlying parameterization of the test case
- Test data
- Expected test result
- Achieved test result
- Measures derived from the tests.

After the application has been put into production, any deviations from regular operation are to be monitored, their causes investigated, and, if necessary, measures for improvement initiated. Indications of significant defects can be, for example, accumulations of deviations from regular operation.

An appropriate procedure for the classification/categorization (protection class) and handling of applications developed or operated by department employees must be defined (Individual Data Processing—IDP). Compliance with programming guidelines is also ensured for the developed IDP applications. Each application is assigned a protection class. If the determined protection requirement exceeds the technical protection possibility of an application, protection measures are taken depending on the results of the protection class classification (BaFin, 2017).

A method established in practice for determining the protection requirement is, for example, that of the Federal Office for Information Security (Federal Office for Information Security, 2017).

The requirements for identifying all applications developed or operated by department employees, for documentation, programming guidelines, testing methodology, determination of protection requirements, and the recertification process of permissions must be regulated (e.g., in an IDP guideline). For an overview and to avoid redundancies, a central register for applications is maintained and at least the following information is collected:

- Name and purpose of the application
- Versioning, date
- Third-party or in-house development
- Responsible department employee
- Technically responsible employee
- Technology
- Result of risk classification/protection requirement classification and possibly
- the derived protection measures.

> **Important** All the aforementioned requirements must be ensured without exception for the implementation and operation of RPA solutions. However, the decisions to be made and the aspects to be regulated in parallel provide operational added value in that they contribute to the lasting success of RPA use.

In practice, a two-phase project approach is recommended (which does not contradict the model described in Chap. 2, but rather provides a process-oriented framework for the application of the model described there):

The first phase aims to anchor the newly introduced technologies in the bank and create the necessary technical framework. This includes the audit-proof setup of the IT infrastructure. In addition to the technical prerequisites, operational anchoring in the bank must be ensured. Here, the action fields "organizational guidelines", "processes", "personnel", and "control" are to be filled with life in an operating model and an adequate governance is to be created. This ensures that all regulatory requirements are met and that ongoing operation and the automation of new processes are organized. Finally, the employees are enabled to use the RPA technology.

In the second phase, the process level is then considered. First, an analysis of existing processes is carried out against the background of potential standardization and automation potentials. Subsequently, a corresponding solution and implementation concept for the optimization of the processes is created, followed by the technical realization and the professional examination. Finally, the rollout and bank-wide communication take place, which also includes the creation of acceptance throughout the institute and the consolidation of the new solution. In this context, the audit plays a key role as an essential part of the qualitative banking supervision in monitoring and evaluating the fulfillment of these requirements (Tomani, 2005).

Fig. 4.8 summarizes the described procedure.

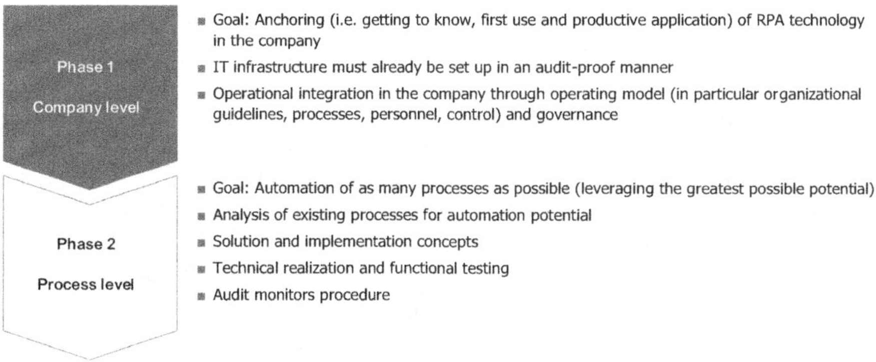

Fig. 4.8 Procedure for RPA use at the higher level. (Own representation based on Tomani, 2005)

4.3.3.3.6 IT Operations

IT operations must meet the requirements resulting from the implementation of the business strategy and from the IT-supported business processes (cf. AT 7.2 Tz. 1 and Tz. 2 MaRisk).

The components of the IT systems and their relationships to each other must be managed in an appropriate manner and the inventory data recorded for this purpose must be updated regularly and on an event-related basis.

The portfolio of IT systems requires control. IT systems should be updated regularly. Risks from outdated or no longer manufacturer-supported IT systems must be managed (life cycle management).

The processes for changing IT systems must be designed and implemented depending on the type, scope, complexity, and risk content. This also applies to new or replacement purchases of IT systems, as well as to security-relevant improvements (security patches).

Changes to IT systems must be recorded, documented, evaluated considering possible implementation risks, prioritized, approved, and implemented in a coordinated and secure manner. Suitable processes must also be established for time-critical changes to IT systems.

Reports on unplanned deviations from regular operation (disruptions) and their causes must be recorded, evaluated, prioritized in particular with regard to possible resulting risks, and escalated according to defined criteria. Standard procedures for measures and communication as well as responsibilities (e.g., for malware on end devices, malfunctions) must be defined. Processing, root cause analysis, and solution finding including tracking must be documented. An orderly process for analyzing possible correlations of disruptions and their causes must be in place. The processing status of open reports on disruptions, as well as the appropriateness of the evaluation and prioritization, must be monitored and controlled. The institute must define suitable criteria for informing the parties involved (e.g., management, responsible supervisory authority) about disruptions.

The specifications for the data backup procedures (excluding data archiving) must be regulated in writing in a data backup concept. The requirements for the availability, readability, and timeliness of customer and business data, as well as the IT systems necessary for their processing, as presented in the data backup concept, must be derived from the requirements of the business processes and the business continuity plans. The procedures for recovery and ensuring the readability of the data must be tested regularly, at least annually, as part of a random sample and on an event-related basis.

The current performance and capacity requirements of the IT systems must be determined. The future performance and capacity requirements must be estimated. Performance delivery must be planned and monitored to identify bottlenecks in a timely manner and respond appropriately. When planning, the performance and capacity requirements of information security measures must be taken into account (BaFin, 2017).

A practical example should illustrate the RPA relevance of this regulatory canon:

Example

What happens, for example, when updates change the interface of the software to be operated? A change in the user interface can mean that the robot needs to be adjusted. Stability and reliability are basic prerequisites for every robot for the rollout and ongoing operation. These prerequisites must be ensured from the outset by proper integration into the company's IT operations and its so-called "change management". ◄

4.3.3.3.7 Outsourcing and other external procurement of IT services

IT services include all forms of IT procurement; this includes in particular the provision of IT systems, projects/trades or personnel provision. The outsourcing of IT services must meet the requirements according to AT 9 of MaRisk (BaFin, 2021). This also applies to outsourcing of IT services that are provided to the institute as in the case of RPA by a service company over a network (e.g., computing power, storage space, platforms or software) and whose offer, use and billing are dynamically adapted to the demand via defined technical interfaces and protocols (cloud services).

► **Important** The institute must also observe the general requirements for the proper organization of business operations in accordance with § 25a para. 1 KWG when procuring IT services from external sources in connection with RPA (see AT 9 para. 1—explanations—MaRisk).

Any procurement of software must be appropriately assessed for the associated risks (see AT 7.2 para. 4 sentence 2 MaRisk).

Due to the fundamental importance of IT for the institute, a risk assessment must also be carried out in advance for any other external procurement of IT services. The other external procurement of IT services must be controlled in line with the strategies, taking into account the institute's risk assessment. The provision of the service owed by the service provider must be monitored in accordance with the risk assessment. The measures derived from the risk assessment for the other external procurement of IT services must be appropriately taken into account in the contract design. The results of the risk assessment must be taken into account in an appropriate manner in the management process of operational risk, especially in the area of the overall risk assessment of the particularly relevant operational risk.

The risk assessments in relation to the other external procurement of IT services must be regularly reviewed and adjusted as necessary, including the contract contents (Becker, 2018).

4.3.3.3.8 IT Emergency Management

The institute must define objectives for emergency management and derive an emergency management process from this. Provision must be made for emergencies in time-critical activities and processes (emergency concept). The measures defined in the emergency concept must be suitable for reducing the extent of possible damage (see AT 7.3 para. 1 MaRisk). The emergency concept must include business continuity and recovery plans. In the case of outsourcing time-critical activities and processes, the outsourcing institute and the outsourcing company must have coordinated emergency concepts (see AT 7.3 para. 2 MaRisk). The effectiveness and appropriateness of the emergency concept must be regularly reviewed. For time-critical activities and processes, it must be demonstrated at least annually and as required for all relevant scenarios (see AT 7.3 para. 3 MaRisk, Berndt et al., 2018).

The objectives and framework conditions of IT emergency management are to be defined on the basis of the objectives of emergency management. The institute must create IT emergency plans based on the emergency concept for IT systems that support time-critical activities and processes.

The effectiveness of the IT emergency plans is to be checked by at least annual IT emergency tests. The tests must fully cover IT systems that support time-critical activities and processes. Dependencies between IT systems or of jointly used IT systems must be appropriately taken into account. An IT test concept must be created for this (BaFin, 2017).

> **Important** Again, it should be noted that it might be beneficial in a first step to limit the use of RPA to non-time-critical activities through an appropriate strategic decision. After all, a robot is not capable of making unprepared decisions.

4.3.3.3.9 Critical Infrastructures

The related supervisory requirements—in context with the other chapters of the BAIT and the other relevant banking supervisory requirements regarding the assurance of appropriate precautions to ensure availability, integrity, authenticity and confidentiality of information processing—specifically to the operators of critical infrastructures (KRITIS operators1).

The regulation in the BAIT thus supplements the banking supervisory requirements for IT with requirements for the effective implementation of special measures to achieve the KRITIS protection objective. The following refers to the preservation of the society's supply security with the critical services mentioned in § 7 BSI-Kritisverordnung (cash supply, card-based payment transactions, conventional payment transactions as well as clearing and settlement of securities and derivative transactions), as their failure or impairment could lead to significant supply shortages or threats to public safety. For critical services, the respective KRITIS operators (and in the case of outsourcing, additionally their IT service providers) must describe and effectively implement appropriate measures that reduce the risks for the safe operation of critical infrastructures to a level appropriate to the KRITIS protection objective. To this end, the KRITIS operators and their IT service providers must adhere to the relevant standards and consider high-availability concepts. The state of the art should be maintained. This chapter can optionally be used to provide evidence according to § 8a para. 3 BSIG in the context of an annual audit. For this purpose, all information technology systems, components or processes of the critical infrastructures must be completely covered in the audit.

Alternatively, the KRITIS operators can pursue a company-specific approach or create an industry-specific security standard (B3S) according to § 8a para. 2 BSIG. The proof according to § 8a para. 3 BSIG is to be provided in these cases with the involvement of a suitable auditing body (see relevant FAQ on the BSI website) (BaFin, 2017).

The scope of the critical infrastructures within the information network must be clearly marked. All relevant interfaces must be included.

All relevant requirements of the BAIT and the other supervisory requirements must be comprehensibly applied to all components and areas of the critical service.

Critical services must be appropriately monitored. Possible impacts of security incidents, also on the critical services, must be assessed.

Within the framework of information risk and information security management according to the BAIT Chaps. 3 and 4, the KRITIS protection objective must be observed and measures for its compliance effectively implemented. In particular, risks that can significantly impair the critical services must be reduced to a level appropriate to the KRITIS protection objective by appropriate risk reduction or avoidance measures. In particular, those measures are suitable which can counteract the risks for availability in case of high and very high protection requirements. Among other things, high-availability concepts should therefore be examined and, if suitable, applied.

The KRITIS protection objective must be taken into account from the determination of the protection requirement through the definition of appropriate measures to the effective implementation of these measures, including the implementation and regular testing of corresponding emergency preparedness measures.

The proof according to § 8a para. 3 BSIG regarding compliance with the requirements according to § 8a para. 1 BSIG can be provided in the context of the annual audit. The KRITIS operator must submit the relevant proof documents to the BSI in due time, in accordance with the currently valid BSI requirements (Becker, 2018).

> **Important** The RPA user is therefore obliged to obtain the necessary proof documents from the selected service provider in good time, to check these in detail and then, if necessary, to submit them to the responsible authorities.

4.3.3.4 Requirements for Models as well as Technology-Supported Innovations and Artificial Intelligence

As part of the upcoming MaRisk amendment 2022, BaFin plans to implement new requirements for models. To this end, a new subsection AT 4.3.5 "Models" is to be added in section AT 4.3 "Internal Control System". With the planned project, BaFin is implementing, among other things, sections 4.3.3 "Technology-Supported Innovations for Lending" and 4.3.4 "Models for Creditworthiness Assessment and for Credit Decisions" of the EBA Guidelines on loan origination and monitoring (European Banking Authority, 2020).

The declared goal is to create a set of rules for all models in the context of MaRisk, although the specific requirements from AT 4.1, AT 4.3.2 as well as BTR 2.1 and BTR 3.1 are likely to remain unchanged in the future.

With the supplementary publication of a principles paper for the use of algorithms in decision-making processes, BaFin provides financial institutions with insight into preliminary considerations on possible minimum requirements for the use of artificial intelligence (AI). Since AI and RPA-based automations are very often used in combination, a consideration of the related requirements of the supervisory authority is appropriate.

In the introduction to the principles paper, BaFin first attempts to define the term artificial intelligence and finds that at the present time, a clear distinction from existing classical statistical methods is not possible.

BaFin understands artificial intelligence as the symbiosis of large amounts of data, high computing power, and machine learning. Machine learning is an umbrella term for data-driven learning methods. This involves differentiating according to the type of algorithms, result types (classification, regression, and clustering), and data types (text, speech, image data) (BaFin, 2021a).

Since machine learning relies on statistical models that are already used in the form of classical regressions in financial institutions, the principles defined in this paper generally apply to all types of decision-making processes that are supported by algorithms. This also includes already established statistical decision models.

Therefore, BaFin recognizes a need to distinguish artificial intelligence from these existing methods. For this purpose, three essential characteristics that characterize artificial intelligence have been defined:

- High complexity of the underlying algorithms
- Short recalibration cycles
- High degree of automation.

> **Important** However, this definition still lacks the sharpness of distinction from the statistical regression models already used in banks. Therefore, it is expected that the supervisory authority will refine the definition.

The principles are intended to provide banks with guidance without restricting their technology neutrality:

1. Overarching Principles
In BaFin's view, guidelines for the use of artificial intelligence should be defined by the management. These guidelines should capture both the potential and the limits and risks of the technology, define risk-limiting measures, and expand the existing IT strategies accordingly.

This requires a comprehensive analysis of potential damages caused by both immediate faulty algorithm-based decisions and mutual dependencies between different AI applications. To adequately meet these requirements, the necessary technical understanding of artificial intelligence must be established directly in the board of directors.

In addition to defining responsibility in the board of directors, BaFin emphasizes in the overarching principles the importance of avoiding systematic biases (Bias) and observing existing, legally prohibited differentiations. Compliance with these principles depends crucially on the underlying training data.

Systematic bias refers to the bias of the AI, which is caused by faulty or incomplete data or by inadequate processing. In this case, the computer adopts the conscious or unconscious human bias and derives logical conclusions from it. As a result, groups of people, ethnicities, or minorities can be systematically discriminated against and disadvantaged.

A well-known example of biases caused by artificial intelligence is the preference for white men in job placements at an online retailer. There, the algorithm had concluded from the training data that white men were more successful in the company. However, this result was biased due to the underlying training data. Such discriminations can arise from human prejudices or also from the use of certain characteristics, which are already not allowed to be used for differentiation according to existing laws. Theoretically, these considerations could also be applied to an AI-driven credit granting process.

This example clearly shows how important the selection and preparation of the training data is, while at the same time it shows the limits of artificial intelligence:

The computer not only learns to recognize correlations, it also learns through our prejudices and preferences. To comply with this enormous responsibility that comes with the selection of data, BaFin therefore demands an overarching review process, which should prevent any discrimination in the institution. This leads to institutions developing an even higher interest in high-quality data, and offers the opportunity to uncover previously established discriminations in processes and to make decisions more fairly in the future.

2. Specific Principles for the Development Phase
In addition to the overarching guidelines, principles are formulated for both the development of AI applications and their later use (BaFin, 2021a).

BaFin requires the documentation and storage of training data as well as high data quality and sufficient quantity. Only through this data storage is it possible to subsequently identify data that have led to biases in AI decisions and therefore need to be removed from the training data set in a validation process. This enables the traceability and reproducibility of the AI results.

It should be noted that particularly complex models for AI—such as neural networks—can hardly meet the requirement of traceability. Thus, BaFin restricts financial institutions in the complexity of AI and the associated opportunity through its use.

The introduction of AI solutions requires, among other things, the selection of high-quality training data, the preparation of these training data, the selection of suitable algorithms, and the setting of parameters. Only by trying out different models can a suitable AI be developed. Therefore, BaFin demands the documentation and justification of model selection, model calibration, and model validation, and thus indirectly the examination of several models to identify a suitable model.

3. Specific principles for application
With the help of artificial intelligence, processes can be automated and decisions can be made by the computer. To ensure a high quality of these decisions and to interpret them, regular validation and constant human control are essential. BaFin recognizes this important role and formulates several minimum requirements for it.

The control mechanisms by humans should be ensured through the following activities (minimum requirements):

Due to the use of AI results as input for further AI applications, possible risk aggregations must be considered and recorded. This means that for each AI application, it must be checked whether the input data have already been determined by an AI and what errors can arise from this. This could be the case, for example, if the creditworthiness of a customer was determined by artificial intelligence in an AI-supported credit granting process. In this way, an inaccuracy in determining creditworthiness can lead to further inaccuracies and errors in the granting process.

Regardless of the business criticality of the application, the output from the AI application must be checked for its correctness and interpretability. With such interpretability analyses (e.g., hiring predominantly white men due to prejudices), errors or inaccuracies in the training data can be identified.

Through plausible interpretation, which goes beyond pure release processes, e.g., through a limit process, review processes of the AI application can be triggered early on. For such a limit process, for example, the quota of detected fraud attempts can be used in an application for fraud detection. If this quota changes over time, an interpretability test is initiated. A process adjustment to minimize fraud could explain a decrease in such a quota. However, if no justification for changed results can be found, a review of the entire algorithm, the input data, and the parameters made must be carried out.

In addition, BaFin will in future also demand the establishment of emergency measures in the event of a failure of the AI application for business-critical processes. In banking practice, RPA is currently not used in time-critical processes, otherwise the challenge would be to maintain functioning business continuity concepts within the framework of business continuity management, thus immediately negating the efficiency gain achieved through the use of RPA by maintaining a double structure.

BaFin also outlines further obligations, such as the regular validation of the application and also an ad-hoc validation, for example in the case of frequent decision deviations, new external or internal risks. The concrete implementation of this position paper will be reserved for a future MaRisk amendment, but should nevertheless be considered in implementation decisions today and seems, if not yet, at least very soon achievable.

Banks already meet extensive requirements for the use and control of AI technologies, among other things in the context of risk management, and thus comply with existing regulatory requirements. Another forward-looking possibility would be certification by the supervisory authority or according to standards recognized by the supervisory authority. In comparison to the EBA, BaFin is still very reserved in this respect. Such certifications recognized by the supervisory authority strengthen user confidence in certain technologies on the one hand and ensure basic regulatory requirements on the other. Especially with AI solutions, existing uncertainties could be effectively reduced in this way.

At the supranational level, the EU AI Act proposed by the European Commission in draft form will have a significant steering effect on all companies that use AI. This regulation is intended to stipulate how all AI applications are classified in terms of their risk and to build risk-appropriate regulatory requirements on this basis. In April 2021, the European Commission presented its proposal for a regulation of the European Parliament and the Council on harmonized rules for Artificial Intelligence, also known as the AI Regulation. The draft regulation is now being discussed in the legislative process by the European Parliament and other EU bodies. A reliable forecast for the date of entry into force is not yet possible today, so that national standards can certainly be used as a revision during the transition period.

For example, the Artificial Intelligence Cloud Service Compliance Criteria Catalogue (AIC4) provided by the BSI contains minimum requirements for the secure use of machine learning. Concrete test criteria are specified for this in eight categories. The AI test catalogue of the Fraunhofer IAIS also contains a guide that is already practical today for identifying AI-specific risks in the dimensions of fairness, autonomy and control, transparency, reliability, security and data protection. These two national criteria catalogues are voluntary, but are extremely helpful audit support instruments until the adoption and entry into force of the EU AI Act (see also BSI, 2023; EUR-Lex, 2021 and Fraunhofer, 2023).

4.3.4 Comparable Regulatory Requirements in Other Sectors of the Financial Industry

The regulatory requirements for the use of RPA in insurance companies, investment companies, and payment service providers, particularly the "Insurance Supervisory Requirements for IT", the "Capital Management Supervisory Requirements for IT", and the "Payment Services Supervisory Requirements for IT of Payment and E-Money Institutions", are fundamentally comparable and differ primarily in sector-specific terminology. The previously established regulatory requirements and implementation guidelines can therefore be transferred accordingly.

At this point, we will only explicitly discuss specifics for payment service providers, as the requirements for the design of the relationship between the bank and the payment service provider are defined in Chap. 11 of the BAIT (BaFin, 2021b):

The risk mitigation measures required by § 53 ZAG to control operational and security-related risks therefore also include measures that directly address payment service users for the reduction, especially of fraud risks. For this purpose, an appropriate management of relationships with payment service users must be established.

The bank must establish and implement processes that improve the awareness of payment service users about the security-related risks associated with payment services by supporting and advising payment service users. This particularly affects communication processes to sensitize their own payment service users to risks when using payment services.

Sensitization can take the form of general addresses (information on the website) or individual addresses as needed. The processes are adapted to the specific current risk and threat situation and can differ with respect to individual payment service users. The support and advice offered to payment service users must be kept up to date and adapted to new risk situations. Adjustments must be communicated to the payment service user in an appropriate manner. If the product functionality allows it, the bank should offer the payment service user the option to deactivate individual payment functionalities. Such deactivation could, for example, include a blocking option for foreign transfers outside the SEPA area. Corresponding requests can be submitted online or in writing. If the bank has agreed on upper limits with the payment service user, the payment service user should be given the opportunity to adjust the agreed limits. This could, for example, include adjusting the daily limit for transfers in online banking.

To detect fraudulent or unauthorized use of the payment accounts of the payment service user, the bank must give the payment service user the opportunity to receive notifications about completed and failed transactions. The aim is to enable the payment service user to have appropriate control of the transactions or attempted transactions carried out, so that fraudulent transactions or attempts at fraud can be detected by the user as early as possible. Constant and immediate explicit notification of all transactions and attempted transactions is not required. Fraud detection measures to be carried out by the bank itself remain unaffected by this.

The bank must inform payment service users promptly about updates to security procedures that have an impact on payment service users in relation to the provision of payment services. The specific communication channel is determined by the institution. The payment service user should be given the opportunity to adequately adjust to and prepare for changed processes in order to be able to use the payment services without interruptions if possible.

The bank must support payment service users in relation to all questions, support requests, notifications of irregularities, or all security-related questions regarding payment services. Payment service users must be adequately informed about how they can receive this support. Appropriate communication channels that can be used by all payment service users are established. These can be made known, for example, via the websites, via technical communication channels, or in written communication.

> **Important** The implementation of the aforementioned obligations, and in particular both standardized customer information and advice as well as the identification of unauthorized payment transactions, typically offer a suitable and wide field of application for RPA solutions.

This will be illustrated using the example of so-called phishing prevention:

Example

The entire audit process of an incoming message can be taken over by a bot. It extracts characteristic elements of the message (URL and IP addresses), compares them with suspicious information structures, checks the message header, and can automatically check attachments. In case of suspicion, mail and attachments are automatically checked with the appropriate protection programs or an alarm is displayed. Only after the successful check without complaint is the message then passed on to another process step. ◄

In all possible fields of application that should be coordinated as well as possible between the bank and the payment service provider, the aforementioned regulatory requirements must be implemented.

4.3.5 Special Requirements from the Works Constitution Act (BetrVerfG)

> **Important** The introduction and operation of RPA is subject to co-determination by the works council under certain circumstances (including the extent and consequences of the planned RPA deployment and company size).

The following regulatory contents are particularly relevant as bases for claims:

§ 87 BetrVerfG Co-determination rights, among other things, in the introduction and application of technical devices intended to monitor the behavior or performance of employees

§ 90 BetrVerfG Employer's duty to inform, among other things, about the planning of work processes and workflows, including the use of artificial intelligence

§ 97 BetrVerfG Co-determination in measures for vocational training of employees, if the use of RPA leads to a change in the activities of the affected employees

§ 111 BetrVerfG Duty to inform about planned operational changes (Holzapfel, 2020)

> **Important** Whether the aforementioned regulations are applicable depends on the exact design of the automation. However, it is generally advisable to involve the employee representation in the considerations and planning for the use of RPA at an early stage.

4.3.6 Special Requirements and Opportunities from the GDPR

Data protection law must be observed when RPA processes personal data. Data protection issues can arise, among other things, when decisions are made by process automation that directly affect a person, e.g., through a chatbot. According to Art. 22 GDPR, a data subject has the right not to be subject to a decision based solely on automated processing, which produces legal effects concerning him or her or similarly significantly affects him or her.

> **Important** This prohibition of automated decision-making can oppose the automation of a process. Systems that automatically reject contracts if certain parameters are not met are prohibited. However, this prohibition is not as strict as it seems at first glance. The law allows various ways in which such a process could be legally designed.

RPA applications that are used in the so-called Smart-Data procedures are also challenging. Recently, the warning letter from the State Commissioner for Data Protection Lower Saxony dated 02.09.2022 (State Commissioner for Data

Protection Lower Saxony, 2022) to cooperative banks in Lower Saxony and the judgment of the ECJ of August 1, 2022, C184/20 (European Court of Justice, 2022) caused significant uncertainty. Both sources concern the application of Smart Data and data processing procedures of personal data in connection with the promotional processing of customer data, as far as they are based on the legal bases in Article 6 paragraph 1 a and f GDPR. The legal basis is already mentioned in the processing directory of the respective responsible party. For procedures based on Article 6 paragraph 1 b to e GDPR and not on a notice or consent solution (e.g., processing for the performance of a contract such as a credit analysis or fulfillment of legal obligations), the developments here are irrelevant. In particular, internal control instruments of the banks are not necessarily affected. However, an RPA deployment for automated and systematized generation of sales opportunities is affected. This has so far been considered permissible by the top associations based on a so-called "notice solution" (customer is informed about the possibility of processing and the compliance with data protection regulations is evaluated by an expert as part of an internal data protection impact assessment). According to the trend at supervisory authorities and courts, data processing via interest balancing in the notice solution will probably be further restricted in the future, so that an efficient use will probably only be possible in the future based on an explicit consent solution. The existing consent declaration had to be further developed for this. In addition to the clarification, to the non-purpose-oriented processing of special categories of personal data, requirements for future data use were also added. The supervisory authorities also demanded some clarifications regarding the increase in transparency and information, which were also incorporated into the development.

In order to simultaneously meet the demand for higher granularity (specificity of processing and processors) in consent, the new consent is also to be optimized so that the customer can decide separately about the use of individual data categories (e.g., payment transaction data, online usage data, and external data). New purposes of use or data categories should be able to be added in the future (e.g., analysis of data according to Art. 9 DSGVO) through modular expandability. The new consent thus creates, based on a current assessment, even higher legal certainty and already covers further requirements for planned use cases. The new consent form is to be implemented with high priority and made available promptly, and customers are then to be asked for their consent. The collection of these consent forms from all customers themselves then represents an exceptionally suitable area of application for corresponding RPA modules. This also applies to ensuring data protection requirements for customers, such as the citizen's right to access stored data, the citizen's right to request electronic copies of all stored data, the reporting of data protection violations, and automated data deletion. All these processes are effective fields of application for RPA, provided that data protection regulations are observed.

Further data protection problems can also arise when RPA is used to monitor the behavior of employees, or when this possibility exists. Prior to a larger-scale RPA deployment, a so-called process mining often analyzes and optimizes company processes. With the help of process mining, business processes are systematically recorded, analyzed, and evaluated. On this basis, the company can assess, correct, optimize, or recreate its processes. Employee data protection aspects and labor law problems arise when process mining or later RPA deployment at the so-called human-machine-human interfaces can prove that individual employees do not adhere to the defined process flows or—compared to the rest of the employees—their work performance drops. If a tool is introduced that makes such behavior control possible, the works council has a right of co-determination. This right of co-determination already exists if the technical facility to be introduced offers the possibility of monitoring the behavior of employees. It is irrelevant whether the company wants to use the technical facility for this purpose at all—it only matters that the possibility exists to be able to control the employees. Resistance usually arises when a company introduces RPA. This introduction is accompanied by the works council's fear that employees will be cut back when the company processes are optimized and automated (Reinkemeyer, 2020). In addition, the legal requirements described in Sect. 4.3.5 must also be taken into account.

In addition to the co-determination rights of the works council, data protection regulations must also be observed when employee-specific information is collected through process mining. The data protection problem can be avoided if the information is collected anonymously and—even retrospectively—the information can no longer be assigned to a specific employee, e.g., if the information always refers to groups of employees. Whether and to what extent this is possible, of course, depends on the company processes that are analyzed and the goals that are pursued with it.

▶ **Important** To nip concerns in the bud, it is imperative to ensure that data access and data transfer are regulated and secured, that planned data deletion is guaranteed, and that the use of RPA is fully integrated into the bank's data protection strategy even before its introduction.

4.3.7 Special legal requirements for IT security

As a special law with effects beyond the financial services sector and implications for the use of RPA, the so-called IT Security Act 2.0 (Second Act to Increase the Security of Information Technology Systems) is particularly relevant here. After being signed by the Federal President and published in the Federal Law Gazette, the law came into force on May 28, 2021. With the IT Security Act 2.0, the first law to increase the security of information technology systems was revised to increase cyber and information security against the backdrop of increasingly frequent and complex cyber attacks and the further advancing digitization of everyday life.

In particular, the numerous changes to Germany's central IT security law—the Act on the Federal Office for Information Security ("BSIG")—are relevant not only for the KRITIS operators already covered by the BSIG due to the tightened IT security obligations and increased fines, but also for a number of other companies. This particularly affects the manufacturers of corresponding IT products that are used in critical infrastructures (and thus also banks) in relation to RPA.

The IT Security Act 2.0 supplements the obligations already existing under the BSIG and introduces new ones (Federal Office for Information Security, 2021):

Obligation to register a Critical Infrastructure with the Federal Office for Information Security ("BSI")
In addition to the existing obligation for operators of Critical Infrastructures to designate a contact point that can be reached at all times for the Critical Infrastructure they operate, a direct obligation to register a Critical Infrastructure is established.

Obligation to Use Attack Detection Systems
The obligation of operators of critical infrastructures, anchored in § 8a BSIG, to take appropriate organizational and technical precautions that are essential for the functionality of the critical infrastructures they operate, is specified. This obligation now also explicitly includes the use of attack detection systems, which must comply with the state of the art.

Obligation to Submit Documents Necessary for an Assessment from the BSI's Perspective and to Provide Information
In connection with this new obligation, the BSI can, for example, request information on key figures regarding the respective threshold values if facts justify the assumption that an operator has not fulfilled his obligation to register.

Obligation to Release Information Necessary to Manage the Disruption
During a significant disruption, the BSI, in agreement with the respective federal supervisory authority, can demand from the affected operators of critical infrastructures or the companies, in the special public interest, the release of information necessary to manage the disruption, including personal data.

Obligations in Connection with the Use of Critical Components
The operator of a critical infrastructure is also imposed with obligations in connection with the use of critical components.

This includes the obligation to notify the Federal Ministry of the Interior, for Construction and Home Affairs ("BMI") of the planned initial use of a critical component before its deployment.

It also involves the operator of a critical infrastructure's obligation to obtain a declaration of trustworthiness (so-called guarantee declaration) from the manufacturer of the critical components. Only after obtaining such a guarantee declaration may the operator of a critical infrastructure deploy critical components. This declaration must be attached to the notification to the BMI.

Based on the above-mentioned notification and the guarantee declaration, the BMI conducts an ex-ante and an ex-post examination regarding the use of critical components and can prohibit the planned initial or further use of a critical component to the operator of the critical infrastructure in agreement with the departments listed in the BSIG and the Foreign Office, or issue orders, "if the (further) use is likely to affect the public order or security of the Federal Republic of Germany". It should be noted that the prohibition of further use of a critical component of a manufacturer can have further consequences for the manufacturer.

In accordance with the above-described obligation of operators of critical infrastructures to use critical components only from manufacturers who have given a declaration of their trustworthiness to the operator of the critical infrastructure, the manufacturers will (have to) give corresponding guarantee declarations throughout the entire supply chain to the operator of the critical infrastructure.

The catalogue of fine provisions has been completely revised: The fine facts have been specified for better enforcement, especially of information and proof obligations, and significantly expanded in accordance with the newly introduced obligations described above. For example, the administrative offence for those KRITIS operators who do not ensure that the contact point to be named is always reachable or—if classified as a company of special public interest according to § 2 para. 14 sentence 1 no. 1 and 2 BSIG—do not submit a self-declaration correctly, completely or in time.

The fines themselves have been drastically increased in order to—as stated in the explanatory memorandum—be able to achieve a steering effect. Instead of the fines of up to 100,000 € or up to 50,000 € possible under the previous BSIG, administrative offences can now—depending on the case—be punished with a fine of (i) up to 2,000,000 €, (ii) up to 1,000,000 €, (iii) up to 500,000 € or (iv) up to 100,000 €.

> **Important** Since in the banking sector the cash supply, card-based and conventional payment transactions and the settlement of securities and derivatives transactions are considered critical by law, a declaration of trustworthiness should be obtained from the involved RPA service providers as a precaution in any case.

4.3.8 Special Requirements of Licensing Law

Through RPA, interfaces can be replaced and data can be exchanged automatically between two systems. For example, the bot receives incoming messages from the mail server, processes them, and forwards them for further processing to another software. If the bot takes over tasks that were previously performed by employees, the licensing law of the software used may be affected. For example, if previously five employees made the input in the respective software and now the bot takes over this work, from the licensor's point of view, this could be considered "indirect software use", which may have been explicitly excluded in the license terms.

It is also possible to automatically trigger certain actions in a software. Because the RPA tool and the software are two different computer programs, the question arises whether a license is required for these automated processes. In this constellation, the use of RPA shows parallels to the so-called "indirect software use" (Witt, 2018).

If the task that is replaced by the RPA tool was previously performed by a human, this person usually needed a license to use the software. If the human is now replaced by a bot, the question arises whether the bot also needs a license. At least the software provider will affirm this. Otherwise, his business model is seriously questioned. This becomes clear when you imagine that the bot not only replaces one human—and thus one license -, but a multitude of humans—and thus a multitude of licenses.

Many software providers therefore prohibit this so-called pooling or multiplexing in their license terms. On the other hand, the company that uses RPA wants to save costs. Additional license costs or even a legal dispute with the software provider pose a risk to the company and the success of the RPA strategy.

Whether and under what conditions indirect use of software requires an additional right of use is currently not clarified. The legal discussion is complex. The legal assessment depends, among other things, on how the computer programs used interact with each other in technical terms. To the author's knowledge, there are no court rulings in Germany that set a trend.

> **Important** Against this background, the affected company should always thoroughly investigate in advance whether there is a risk at all with regard to the software used and the underlying license model that a discussion about indirect use must be conducted.

If the licensor has, for example, explicitly regulated whether and how indirect use is possible, RPA can be introduced if these conditions are met and implemented. In case of doubt, it should be legally checked whether there could be a problem here.

4.4 Conclusion and Recommendation for Action: Dealing with RPA in the Context of Regulatory Framework Conditions

RPA can undoubtedly contribute to solving the diverse and pressing challenges facing the financial services sector.

> **Important** The use of RPA is highly regulated by the legislator and the supervisory authority, as shown, from many different perspectives. However, these regulatory requirements are not new and, with the exception of the requirements for the use of AI, are not specific to RPA and therefore do not present the financial industry with insurmountable challenges and can all be met with careful and systematic planning.

Such a structured and granular approach is not only a regulatory end in itself, but also undoubtedly contributes to the overall economic success of the implementation measure(s).

In addition to the clear and early strategic decision-making by the company management and the resulting creation of a stringent strategic "mindset" among the employees of the institute, it is particularly the various risk analyses required by supervisory law (with regard to AT 8 and possibly AT 9 MaRisk, sections 3.7, 3.9, 3.11 and 8.5 of the BAIT), the careful and well-founded preparation of which creates the decisive foundations for the later implementation success of RPA. In parallel, the foundations for the non-trivial documentation requirements associated with the use of RPA are also created. The supervisory challenges, which are sometimes completely wrongly perceived as a necessary evil, have a well-structuring effect on the introduction and later operating process when approached and implemented systematically. The mandatory involvement of various departments of the institute and the various function holders of the first, second and third lines of defense also ensures that a sufficiently large number of personnel capacities are familiar with the topic and that risk concentrations do not even establish themselves in this regard.

Even the multidimensional, further legal requirements highlighted in the previous chapter, which need to be considered and observed in the introduction and operation of RPA, are introduced to the user through careful execution of the risk analysis(es) required by supervisory law, thereby indirectly making a major contribution to an initially and permanently successful use of RPA.

- RPA (also paired with other technology enhancements such as AI etc.) can generate competitive advantages and is increasingly being used in banks and financial service providers
- This results in the necessity to examine the technology, its application scenarios and its use from the perspective of the audit
- Important: RPA can be considered as an audit object (understanding here), but can also serve as a tool for audit support
- The legal basis to be considered in an audit of RPA in banks is particularly the KWG (and especially the MaRisk and the BAIT), but subsequent other legal and regulatory norms such as the BGB or the DSGVO are also relevant
- If the risk analysis(es) required by supervisory law are carried out and a systematic/structured approach is chosen, there is nothing to prevent a permanently successful use of RPA from the perspective of the audit

In the final Chap. 5, the results of the technically oriented contributions in Chaps. 2, 3 and 4 are to be operationalized and transferred into a user-oriented implementation plan.

References

Allweyer, T. (2016). *Robotic Process Automation—Neue Perspektiven für die Prozessautomatisierung* (Working Paper Fachbereich Informatik und Mikrosystemtechnik Hochschule Kaiserslautern). http://www.kurze-prozesse.de/blog/wp-content/uploads/2016/11/Neue-Perspektiven-durch-Robotic-Process-Automation.pdf. Accessed: 5. Sept. 2022.

Atruvia. (2022). *Banking digitale transformation*. https://atruvia.de/leistungen/wir-sind-die-moeglichmacher/banking-digitale-transformation. Accessed: 14. Sept. 2022.

BaFin. (2017). *Rundschreiben 10/2017 (BA), Bankaufsichtliche Anforderungen an die IT (BAIT) vom 03.11.2017.*

BaFin. (2019). *Protokoll zur Sitzung des Fachgremiums MaRisk am 15.03.2018.*

BaFin. (2021). *Rundschreiben 10/2021 (BA)—Mindestanforderungen an das Risikomanagement—MaRisk.*

BaFin. (2021a). *Aufsichtliche Prinzipien für den Einsatz von Algorithmen in Entscheidungsprozessen von Finanzunternehmen, Prinzipienpapier vom 15.06.2021.*

BaFin. (2021b). *Rundschreiben 11/2021 (BA)—Zahlungsdiensteaufsichtliche Anforderungen an die IT (ZAIT), Rundschreiben vom 16.08.2021.*

Becker, A. (2018). *Bearbeitungs- und Prüfungsleitfaden: MaRisk (IT) und BAIT-Mindestanforderungen an den Einsatz von Informationstechnik in Banken.*

Becker, A. (2022). *Prüfungsleitfaden BAIT, Sicherer IT-Einsatz in Kredit- und Finanzdienstleistungsinstituten.*

Behrens. (2021). *RPA in Banken—was ist mit der Compliance?*https://www.movisco.de/rpa-in-banken-compliance. *Accessed: 14. Sept. 2022.*

Berndt, M., et al. (2018). *Neue MaRisk.* Finanz Colloquium Heidelberg.

Berthel, J., & Becker, F. (2003). *Personal-Management* (7th edn.). Schäffer-Poeschel.

Bonertz, et al. (2020). *Prozessautomatisierung*. https://www.nttdata-solutions.com/FC3-Neuer-BPM-Trend-Robitic-Process-Automation.pdf. Accessed: 11. Sept. 2022.

Brettschneider, J. (2020). Bewertung der Einsatzpotenziale und Risiken von Robotic Process Automation Evaluation of the Potential Uses and Risks of Robotic Process Automation. *HMD Praxis der Wirtschaftsinformatik, 57,* 1097–1110.

Bretz, J. (2015). *IT im Fokus der Bankenaufsicht-Anforderungen der Aufsicht.* Finanz Colloquium Heidelberg.

Bundesamt für Sicherheit in der Informationstechnik. (2017). *Standard B.S.I 200-2: IT-Grundschutz-Methodik.*

Bundesamt für Sicherheit in der Informationstechnik. (2020). *Kriterienkatalog Cloud Computing C5.*

Bundesamt für Sicherheit in der Informationstechnik. (2021). *Zweites Gesetz zur Erhöhung der Sicherheit informationstechnischer Systeme (IT-Sicherheitsgesetz 2.0).* https://www.bsi.bund.de/DE/Das-BSI/Auftrag/Gesetze-und-Verordnungen/IT-SiG/2-0/it_sig-2-0_node.html. Accessed: 24. July 2022.

Bundesamt für Sicherheit in der Informationstechnik. (2023). *Kriterienkatalog für KI-Cloud-Dienste—AIC4.* https://www.bsi.bund.de/DE/Themen/Unternehmen-und-Organisationen/Informationen-und-Empfehlungen/Kuenstliche-Intelligenz/AIC4/aic4_node.html. Accessed: 14. Febr. 2023.

Bundesverband der Volks- und Raiffeisenbanken. (2020). *Muster-Risikobewertung nach BAIT mit BVR-Muster-Risikoanalyse MaRisk in einer Datei zusammengeführt, Mitgliederinformation vom 06.02.2020.*

BVR. (2020). *BVR-Umfrage zur Automation von Prozessen in Genossenschaftsbanken-Ergebnispräsentation vom 21.09.2020.*

Claaßen, M. (2015). *Anwendungsmanagement unter Berücksichtigung der MaRisk.* Finanz Colloquium Heidelberg.

Conrads, J., et al. (2015). *Benutzerrechte: Vergabe, Protokollierung, Überwachung.* Finanz Colloquium Heidelberg.

Dagianis. (2021). *Sicherer und aufsichtskonformer Einsatz von RPA-Technologie*. https://blogs.pwc.de/de/digital-trust/article/224450/sicherer-und-aufsichtskonformer-einsatz-von-rpa-technologie/. Accessed: 14. Sept. 2022.

Deutsche Bundesbank. (2017). Der aufsichtliche Überprüfungs- und Bewertungsprozess für kleinere Institute und Überlegungen zur Proportionalität. *Monatsbericht Oktober, 2017*, 45–58.

Douqué, et al. (2022). *Intelligente Automatisierung in der Immobilienfinanzierung*. https://bankinghub.de/operations/intelligente-automatisierung-immobilienfinanzierung?utm_medium=email&utm_campaign=BankingHub%20%20Newsletter%20232022&utm_content=BankingHub%20%20Newsletter%20232022+CID_f9804ff316c20af38e534054c0039381&utm_source=CM%20Newsletter&utm_term=Weiterlesen. Accessed: 5. Dec. 2022.

EUR-Lex. (2021). *Laying down harmonised rules on artificial intelligence (Artificial Intelligence Act) and amending certain union legislative acts*. https://eur-lex.europa.eu/legal-content/EN/TXT/?uri=celex%3A52021PC0206. Accessed: 14. Febr. 2023.

Europäische Kommission. (2020). *Entwurfsfassung Digital Operational Resiilience Act (DORA)*.

Europäischer Gerichtshof. (2022). *Vorlage zur Vorabentscheidung vom 01.08.2022—C-184/20*. https://dejure.org/dienste/vernetzung/rechtsprechung?Gericht=EuGH&Datum=01.08.2022&Aktenzeichen=C-184%2F20. Accessed: 1. Nov. 2022.

European Banking Authority. (2014). *Leitlinien zu gemeinsamen Verfahren und Methoden für den aufsichtlichen Überprüfungs- und Bewertungsprozess (SREP)* (EBA/GL/2014/13), London, Titel 6–8.

European Banking Authority. (2019). *Leitlinien zu Auslagerungen* (EBA/GL/2019/02).

European Banking Authority. (2020). *Guidelines on loan origination and monitoring* (EBA/GL/2020/06).

EY. (2018). *Mangelnde Vorbereitung auf die Integration von RPA kann ihren potentiellen Nutzen verringern*. Die Interne Revision (IR) kann helfen, die Risiken zu minimieren. https://www.ey.com/de_de/consulting/how-internal-audit-can-help-make-rpa-implementation-a-success. Accessed: 28. Oct. 2022.

FAZ. (2022). *EZB durchleuchtet mit KI-Werkzeug „Heimdall" Kandidaten für Bankvorstände*. https://www.faz.net/aktuell/finanzen/ezb-durchleuchtet-kandidaten-fuer-bankenvorstaende-18170182.html?xing_share=news. Accessed: 14. July 2022.

Feldmann, C. (2022). *Praxishandbuch Robotic Process Automation (RPA) Von der Prozessanalyse bis zum Betrieb*.

Fraunhofer IAIS. (2023). *KI-Prüfkatalog*. https://www.iais.fraunhofer.de/de/forschung/kuenstliche-intelligenz/ki-pruefkatalog.html. Accessed: 14. Febr. 2023.

Genossenschaftsverband-Verband der Regionen e. V. (2021). *Hinweise zur Anwendung von MaRisk AT 8.2 bei Moduleinführungen*.

Glaser, C. (2019). *Risiko im Management*. Springer Gabler.

Gleißner, W., & Romeike, F. (2005). *Risikomanagement* (pp. 211 ff.). Rudolf Haufe Verlag.

Hanenberg. (2001). Neue Entwicklungen bei bankaufsichtlichen Regelungen zur Internen Revision. *Die Wirtschaftsprüfung, 54*(7), 392–406.

Hannemann, R., Schneider, A., & Hanenberg, L. (2008). *Mindestanforderungen an das Risikomanagement (MaRisk). Eine einführende Kommentierung*. Schaeffer-Poeschel.

Hannemann, R., et al. (2019). *Mindestanforderungen an das Risikomanagement, Kommentar*. Schaeffer-Poeschel.

Helfer, M., et al. (2008). Erfüllung der Anforderungen der MaRisk unter Berücksichtigung von Öffnungsklauseln-Allgemeiner Teil (AT 4-AT 9). In *MaRisk-Öffnungsklauseln: Prüfungsvorbereitende Dokumentation*. Finanz Colloquium Heidelberg.

Hofer, M. (2011). MaRisk-Erneute Überarbeitung vor dem Hintergrund internationaler Standards. *BaFin-Journal, 1*, 6–10.

Holzapfel, M. (2020). https://marco-holzapfel.de/2020/11/11/robotic-process-automation-rpa-und-mitbestimmung-des-betriebsrats. Accessed: 14. Oct. 2022.

Internal Auditing. (2020). *Durch Den Einsatz Von Bots Revisionskapazitäten Vervielfachen.* http://internalauditing.de/w020617p/blog/2020/01/02/durch-den-einsatz-von-bots-revisionskapazitaeten-vervielfachen/. Accessed: 14. Sept. 2022.

Koenen, D. (2016). *Wie Banken mit regulatorischen Vorgaben in der IT umgehen sollten.* Springer Professional.

Kokert, J., & Held, M. (2013). *IT-Sicherheit: Erwartungen der Bankenaufsicht.* https://www.bafin.de/SharedDocs/Veroeffentlichungen/DE/Fachartikel/2013/fa_bj_2013_11_it_sicherheit.html. Accessed: 14. Dec. 2022.

Krekel, C., & Faulmann, B. (2013). Internes Kontrollsystem reduziert Risiken. In *diebank Ausg.* 1/2013, (p. 58).

Kurowski, H. (2004). Risikoorientierte Prüfung der Innenrevision. *Betriebswirtschaftliche Blätter, 53*(9), 470–473.

Landesbeauftragte für Datenschutz Niedersachsen. (2022). *Schreiben an die Genossenschaftsbanken in Niedersachsen vom 02.09.2022.*

Lomanto, D. (2019). *What does it mean to think with an automation first mindset?*https://www.UiPath.com/blog/automation/automation-first-mindset. *Accessed: 16. Sept. 2022.*

Luz, et al. (2011). *Kommentar zum Kreditwesengesetz (KWG) inklusive SolvV, LiQV, GroMiKV, MaRisk.*

Ostrowicz, S. (2017). *Automatisierung von Serviceprozessen.* https://www.horvath-partners.com/de/magazin/ausgabe-022017/automatisierung-von-serviceprozessen. Accessed: 1. Nov. 2022.

parcIT. (2022). *Muster-Risikohandbuch 4.0.*

Petersen, J., & Schröder, H. (2020). Entwicklung einer Robotic Process Automation(RPA)-Governance. In *HMD Praxis der Wirtschaftsinformatik.* (Vol. 57, No. 6, pp. 1130–1149). Springer.

PWC. (2021). *Sicherer und aufsichtskonformer Einsatz von RPA-Technologie.* https://blogs.pwc.de/de/digital-trust/article/224450/sicherer-und-aufsichtskonformer-einsatz-von-rpa-technologie/. Accessed: 9. Nov. 2022.

Reinkemeyer, L. (2020). *Process Mining, RPA, BPM, and DTO.* Springer Professional.

Schmidt, C. (2020). *RPA Strategie—Entwerfen eines Target Operating Models.* https://www.ifb-group.com/blog/de/rpa-strategie-entwerfen-eines-target-operating-models. Accessed: 19. Sept. 2022.

Spateneder, E. (2005). Personalentwicklung unter dem Blickwinkel der MaRisk. In Eller (Ed.), *Gesamtbanksteuerung und qualitatives Aufsichtsrecht* (pp. 89–104). Deutscher Sparkassenverlag.

Tomani, H. (2005). IR als wesentlicher Bestandteil der qualitativen Bankenaufsicht. In Eller (Ed.), *Gesamtbanksteuerung und qualitatives Aufsichtsrecht* (pp. 76–88). Deutscher Sparkassenverlag.

Tsolkas, A., et al. (2010). *Rollen und Berechtigungskonzepte.* Springer Vieweg.

UiPath. (2022). *Glossary.* https://docs.UiPath.com/overview-guide/docs/UiPath-glossary. Accessed: 9. Nov. 2022.

Universität Innsbruck. (2022). *Auswirkungen sich verändernder Wertschöpfungsketten im Finanzsektor auf die IT-Sicherheit.* https://www.bafin.de/SharedDocs/Downloads/DE/Bericht/dl_abschlussbericht_forschungsprojekt_uni_innsbruck.pdf?__blob=publicationFile&v=3. Accessed: 12. Dec. 2022.

Witt, H. (2018). *Braucht der Software-Roboter eine eigene Nutzerlizenz?*https://rpa-journal.org/braucht-der-software-roboter-eine-eigene-nutzerlizenz/. *Accessed: 5. Sept. 2022.*

Wundenberg, M. (2012). Compliance und prinzipiengeleitete Aufsicht über Bankengruppen. In *Schriften zum Unternehmens- und Kapitalmarktrecht* (Vol. 1, pp. 92 ff.). Mohr Siebeck.

5 Implementation Roadmap

After the previous chapters have provided an insight into RPA technology as such, its possible strategic significance, and—most relevantly—the requirements and evaluation from a revision perspective, this chapter aims to consolidate the results. For this purpose, a schematic implementation roadmap is created, which is intended to assist readers (and decision-makers regarding the use of the technology) in making a pro/con decision about RPA use and also in the later implementation steps. Fig. 5.1 shows the "implementation roadmap" for RPA at a glance. This is intended as a thought aid and helps in checking whether the use of RPA in one's own company makes sense and should be implemented. In particular, the perspectives that are in focus within this book are highlighted.

In the first step, the strategic competitive perspective is examined. The fields highlighted in yellow show the details here. The first question to answer is what strategy or strategic direction the institute is pursuing. The strategies included in Fig. 5.1 are to be understood as examples. Thus, the authors do not claim to have included all practice-relevant manifestations. For a successful strategy implementation, it is important that the strategic profile is recognizable to the customer. RPA can help to establish or reinforce this recognizability a) or b). In the case of a desired cost leadership, a (process) cost reduction using RPA helps to operationalize the strategy by achieving price advantages for customers. In the case of a desired differentiation, advantages through new processes that are only feasible thanks to RPA are conceivable, for example.

> **Important** RPA is not always suitable. The technology is—figuratively speaking—one of many tools in the toolbox of a good process manager.

If the basic decision is made to use RPA, operational issues are the first to be addressed. These can be identified by the fields highlighted in green. The entry point is the selection of the areas that should be improved first. This list does not

M. R. Smeets et al., *Robotic Process Automation in Use*, https://doi.org/10.1007/978-3-658-45902-4_5

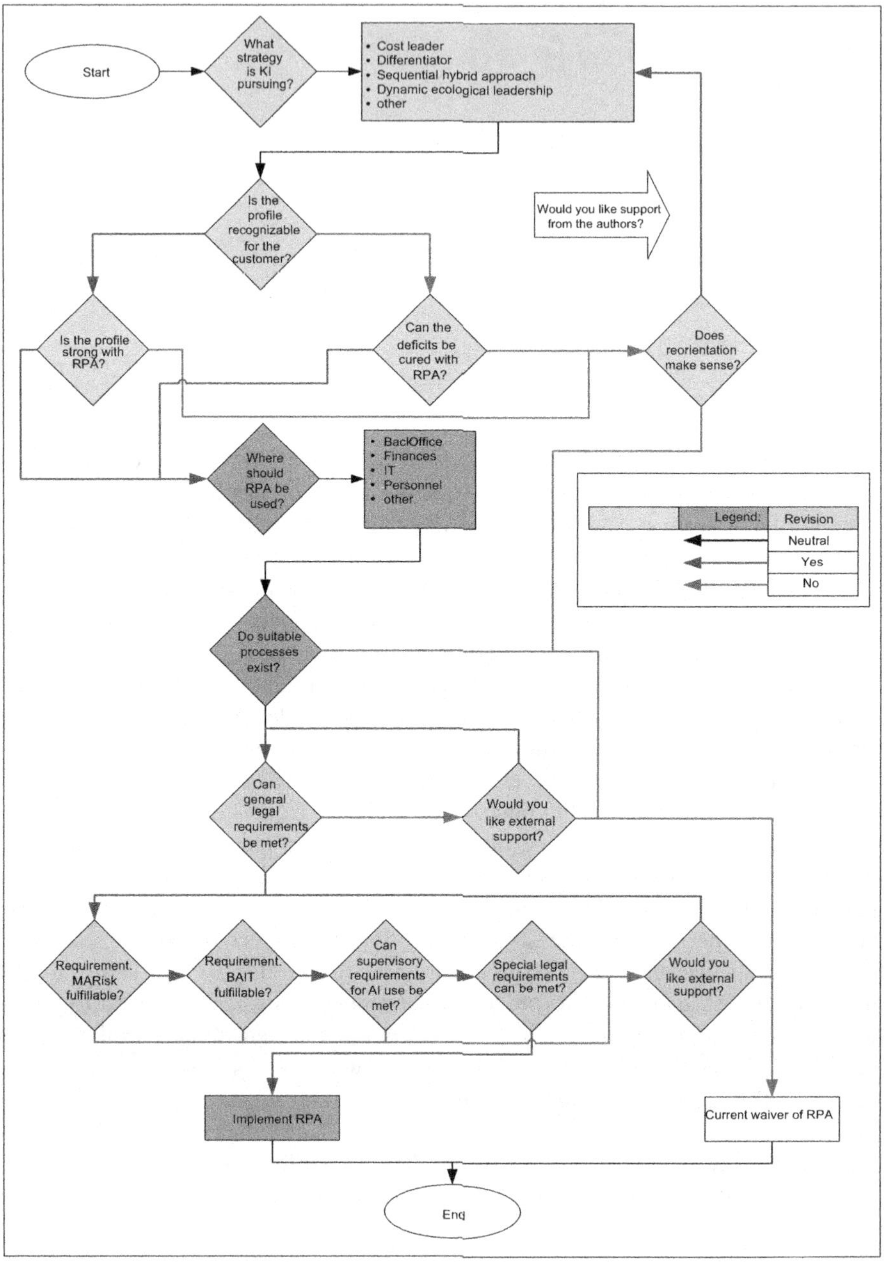

Fig. 5.1 Comprehensive audit process for RPA introduction (own illustration)

claim to cover all (potentially) relevant areas. Subsequently, it must be clarified whether or which of the functional areas contain suitable processes.

After a positive preliminary decision for a specific RPA application, the (supervisory) legal perspective must be checked. The aspects described in this book are relevant in this context. They concern the (supervisory) legal classification of RPA technology, which may and will lead to different assessments depending on the institution and in accordance with the supervisory principle of (double) proportionality, according to which the extent of the respective risk management must be oriented towards the type, scope and complexity of the institution's business. The introduction of RPA is always associated with various technical, structural and procedural change processes. To systematically address these, the financial supervisory authority provides an organizational "blueprint" with the requirement from AT 8.2 MaRisk. Before major changes in the organizational structure and procedures as well as in the IT systems, the institution must analyze the effects of the planned changes on the control procedures and the control intensity. The organizational units later integrated into the workflows must be involved in these analyses. The risk controlling function, the compliance function and the internal audit must also be involved within the scope of their tasks. In this context, it is necessary to meet the general legal requirements as well as the specific requirements for institutions and financial service providers (for banks, these are in particular the MaRisk and the BAIT). The relevant key aspects are highlighted in blue in the process plan. Once the basic preliminary checks to meet the legal framework conditions have been completed with a positive result, the concrete implementation phase follows. Here, the specific RPA process is to be implemented in an (ideally) 8-step process (see Fig. 2.6). With this, the process has reached its end and can—if necessary—start again at a later date.

Made in the USA
Monee, IL
13 September 2026

62973751R00077